Yale Historical Publications

The Hunter's Game

Poachers and Conservationists in Twentieth-Century America

Louis S. Warren

yale university press   new haven and london

Published under the direction of the Department
of History of Yale University with assistance from
the income of the Frederick John Kingsbury
Memorial Fund.

Printed in the United States of America.

Library of Congress
Cataloging-in-Publication Data
Warren, Louis S.
The hunter's game : poachers and conservationists
in twentieth-century America / Louis S. Warren.
    p.   cm.
Includes bibliographical references (p.   )
and index.
ISBN 0-300-06206-0 (cloth : alk. paper)
ISBN 0-300-08086-7 (pbk. : alk. paper)
1. Hunting—United States—History—20th
century.   2. Poaching—United States—History—
20th century.   3. Hunting customs—United
States—History—20th century.   4. Wildlife
conservation—United States—History—20th
century.   5. Wildlife management—United
States—History—20th century.   I. Title.
SK41.W36   1997   97-13847
    95'49'0973—dc21   CIP

A catalogue record for this book is available from
the British Library.

The paper in this book meets the guidelines for
permanence and durability of the Committee on
Production Guidelines for Book Longevity of the
Council on Library Resources.

10 9 8 7 6 5 4 3

For Spring

Contents

Acknowledgments

I have acquired more debts on this project than I can ever repay. The idea for this book developed out of conversations with Bill Cronon while I was a teaching assistant for his environmental history course. Bill has been a consistent mentor and friend; without his help, his unfailing critical eye, his constant encouragement and steadfast faith, the project would not have been possible.

Howard Lamar was another great guide through the labyrinth of research, writing, and publishing. Like so many colleagues across the country, I am convinced that the academic profession has not seen a kinder, more supportive, and more knowledgeable teacher.

Special thanks to Jim Scott. He signed on as a dissertation reader at my request, quite late in the process, and became a wonderful supporter and critical reader of the work.

Others were also very helpful: Garry Brewer was a concerned and motivating advisor and "guardian" during my year at the Institution for Social and Policy Studies; Tim Clark graciously guided me through the rudiments of wildlife ecology at the Yale School of Forestry and Environmental Studies; Bob Harms was mentor, critical reader, and friend. Todd DePastino, Kevin Rozario, and Lane Witt have seen this work develop from idea to book; discussions with them have been nothing short of inspirational.

There were many others whose helpful input has shaped this project, including Sara Deutsch, Johnny Mack Faragher, Mark Fiege, Karl Jacoby, Maria Montoya, Gunther Peck, and Sam Truett. My colleagues at the University of San Diego were constantly encouraging, especially Iris Engstrand, Michael Gonzalez, Jim Gump, and Lisa Cobbs Hoffman. Patty Limerick and Dan Flores offered very useful critiques of the work, and Richard White was a very engaged and extremely helpful reader for Yale University Press. My thanks for all their time and effort.

In the field, there were numerous people who directed me to valuable resources and smoothed my path. Cheryl Bodan at the Pennsylvania Game Commission Library; Paul Roberts and the staff of the Historical Society of Western Pennsylvania; Robert Castelucci and his staff at St. Lawrence Church in Hillsville; Kristin Kahrer at the Lawrence County Court House in New Castle and the staff of the Allegheny County Law Library in Pittsburgh; William Smith at Pinkerton International; Ellis Hoffman and the Lawrence

Acknowledgments

County Historical Society; Beth Dunagan at Glacier National Park's Ruhle Library; the staff of the New Mexico State Records Center and Archives; John Grassham, Rose Diaz, and Michael Miller at the Center for Southwest Research at the University of New Mexico. Bill deBuys was generous with his time, contacts, and insights as I embarked on interviews and research in New Mexico; Bill Hall at the University of San Diego Inter-Library Loan Office never once failed to find and deliver the materials I needed as I went into final revisions.

I would like to offer special thanks to Darrell and Roberta Kipp, who helped me get my feet on the ground with my Montana interviews. I would not have been able to find my way through the complex history of northwestern Montana without them. Mark Spence provided extremely useful insights into the thesis of the book and liberally dispensed his own research sources on the Blackfeet and Glacier National Park. Many people shared life experiences and local lore with me. I would especially like to thank Teddy Burns, Les Davis, Linda Davis, Joe Fisher, Bob Frauson, Merlin Gilham, Gary Hannon, Fred Iovanella, Juanita McKee, Michael Miller, Moises Morales, Louis Perrott, Frank Piscueneri, Alfred Retort, Joe Rich, Clarence Wagner, and Curly Bear Wagner.

A number of institutions supported the research and writing of the work, including the Andrew W. Mellon Foundation, the Western History Association, the American Philosophical Society, the Historical Society of Western Pennsylvania, and the Yale University History Department. Much essential revision occurred under the auspices of the University of San Diego's Faculty Research Grants, which provided me with release time to complete the manuscript.

Some material in chapter 1 appeared previously in *Pittsburgh History* (1991–92), published by the Historical Society of Western Pennsylvania. Some material in chapters 5 and 6 appeared previously in *Transactions of the Sixtieth North American Wildlife and Natural Resources Conference* (1996), published by the Wildlife Management Institute; used with permission.

There has been discussion in academic circles of the need to move degree candidates to completion more quickly. The best way to accomplish that would be to give every graduate student parents like mine. Throughout this endeavor, I have been the beneficiary of my parents' unstinting moral support as well as occasional, much-needed grants from the Claude and Elizabeth Warren Bank of Higher Education and Field Research. To them, a heartfelt thanks.

My eldest son was as encouraging as my parents. At the age of six, Jesse advised me to take my time on "that work," so that I would "do it right the first time." I took my time, and not a little of his—he's now thirteen. I wish to thank him for his extraordinary assistance, not least in the care and maintenance of his little brother, Sam, who joined us three years ago. No writer could ask for greater inspiration than these boys.

I owe the most to my wife, Spring. For the past nine years, I have had the constant advantage of her "slash and burn" editorial skills, her invaluable insights into western life and history, and her sense of humor. Knowing the West as she does, her help in directing my inquiries, refining questions, locating sources, and writing complicated stories has made it far easier to complete the work than would have been possible otherwise. More important, she has been friend, co-conspirator, and co-adventurer in all the years of moving back and forth between New Mexico and Montana, New Haven and points west. To her, my greatest fan and toughest critic, this work is affectionately dedicated.

The Hunter's Game

Prologue
Going West
Wildlife, Frontier, and the Commons

Let me go to the Blue Land
Where the deer fawns are born
And the clouds hover over them
And the rain descends.
Let me go to the Yellow Land
Where the antelope fawns are born
And the clouds hover over them
And the rain descends
And the rivers begin to flow.
—Hunter's Song, Santa Ana Pueblo

Wild animals would not stay in a country where there were so many people. Pa did not like to stay either. He liked a country where the wild animals lived without being afraid. He liked to see the little fawns and their mothers looking at him from the shadowy woods, and the fat, lazy bears eating berries in the wild-berry patches. In the long winter evenings he talked to Ma about the western country. In the West the land was level, and there were no trees. The grass grew thick and high. There the wild animals wandered and fed as though they were in a pasture that stretched much farther than a man could see, and there were no settlers. Only Indians lived there.—Laura Ingalls Wilder

For Ben Senowin and his companions, July 13, 1895, was a terrifying day. They awoke to find that twenty-seven armed white men had surrounded their camp during the night. In another era this might have been expected of whites. But not now. Senowin and his friends were Bannock Indians, from the Fort Hall Reservation in Idaho. They had hunted elk in Jackson Hole, Wyoming, for decades, and in their treaty with the whites, signed a generation before, they were guaranteed the right to hunt elk on unclaimed public lands. They had pitched their tipis on the banks of the Fall River, in the remote northwestern corner of Wyoming. They avoided white settlements and homesteads and hunted the open spaces, expecting to gather only enough elk meat and hides to see them through winter.

But the white posse, under the command of one Constable Manning, threatened to kill Senowin and the others unless they handed over their property. The whites took everything: nine tipis, twenty horses, with saddles and blankets, seven rifles with ammunition, and all the Bannocks' elk meat. Then they arrested Senowin and his party. The charge: violating the game laws of the state of Wyoming.

The posse separated out the men and placed them under armed guard. They took the women and children fifty yards away, where armed guards surrounded them. Then, with the Indians at gunpoint, and Senowin and the eight other Bannock men in front, they began a journey toward Marysville, fifty-five miles away. They traveled thirty miles, until it was almost dark.[1]

Federal officials later reported that the arrests were part of the settlers' plan to bring the issue of Indian hunting rights before the courts. Bannocks may have hunted the area for years, but in the 1880s and early 1890s white immigrants had filled much once-empty land with small ranches and farms. The Bannocks steered well clear of these settlements. But many local whites made sizable profits in the fall and early winter as hunting guides for "dudes," tourists from eastern states and Europe who paid cash for the chance to hunt elk. To the guides of Jackson Hole, Indian elk hunts seemed a needless drain on elk populations, which were increasingly a marketable resource. The issue of Indian hunting, and what to do about it, had dominated local elections in the fall of 1894. In the end, the settlers hatched a plot to instigate a fight with the Indians, even if it meant killing some of them, in hopes of gaining federal assistance to push the Indians out of the hunting grounds at Jackson Hole.

But the Bannocks did not know all this. They were frightened and confused. They did not know where they were going, or why these men should object to their hunting expedition. As evening came on, they approached a dense wood. Constable Manning spoke to his men, who then loaded cartridges into their weapons. At the rear of the procession, the women and children saw this and began to wail, anticipating that the men would be killed. Senowin passed word to the men: when we reach the woods, run. And they did, making a break for the trees into gathering darkness, a twilight gray suddenly ripped by the blaze of guns.

The women and children scattered in all directions as two men fell. The rest of the Bannocks reached the trees, where they hid and waited for morning.

They regrouped then, and returned to the scene after making sure that the whites were gone. Two babies were missing. So were the two men who had fallen in the burst of rifle fire: Nemuts, twenty, and Tanega, an old man. Unable to find the missing people, Senowin and the others returned to Fort Hall, luckily meeting other Indians, who gave them food along the way.

Of the four who were missing, two returned. Nemuts was shot through the back, the bullet lodging in his forearm. He lay still that evening, feigning death until the posse left. He then crawled several miles and survived seventeen days in the wild. A passing party of Mormons found one of the infants and took him to Fort Washakie. The other child was never found. Government scouts found the body of Tanega days later. The old man had been shot four times in the back.

In many ways this story resembles the most bitter chapters in western history, the confrontations between Indians and whites that spanned generations. Like them, it contains bloodshed, tragedy, and not a little treachery. But if this incident seems like a sad coda to the Indian wars, it also contains hints of something different. On one level, it was whites attacking Indians. But on another, it was a party of state agents enforcing state game laws against local hunters. To be sure, there were locals on both sides at Fall River, for the posse was comprised of local whites. But the posse was part of the state's legal apparatus for securing control over the traditional Bannock hunting grounds. To the Bannock, Jackson Hole had long been a communal hunting area, a local commons.[2] The posse represented the vanguard of a new order, state regulation of fish and game. If we pay close attention to the confrontation at Fall River, and to its dimensions as a struggle between local community and state authority, we can begin to appreciate its deeper meanings for western history.

If it seems a typical story of the West in its dramatic conflict between Indians and whites, it might also seem a classic western tale by virtue of the prominent role of wild animals: it was the elk of Jackson Hole around which the bloody and sad events of that summer swirled. Elk brought the Bannocks on their long trip from Fort Hall. Because elk also drew eastern and European hunters to Jackson Hole, they provided white settlers with a way to earn cash. Settlers were willing to kill for elk; and Bannocks would die for them.

Perhaps as often as they have been about Indian-white conflicts, stories of the West have been stories about wild animals. In our own era and in centuries past, "west" has been where many Americans have gone for wildlife. The "dudes" who went to Jackson Hole to hunt were in a sense following a venerable tradition. Whether it was Daniel Boone in the eighteenth century, Buffalo Bill in the nineteenth century, or tourists looking for buffalo and elk in twentieth-century national parks, the West has exerted a powerful hold on American imaginations as the place where "the buffalo roam, where the deer and the antelope play." Visions of bountiful wildlife have been a lure of the land, and as such, close to the heart of America's westering experience.

Wildlife was and is as much symbol as substance. In American popular culture, the buffalo stampede has long assumed a significance beyond the animals themselves. Allusions to huge herds of the animals thundering across the plains evoke abundance and freedom. To many settlers in the nineteenth century, plentiful game meant that the land would support domestic animals, and possibly crops. This was pure folklore, but not totally at odds with scientific facts. Completely dependent on local plants and natural features for their food and shelter, wild animals exist only in a larger context of local ecosystems. In the words of one wildlife biologist, "we cannot say exactly where the animal ends and the environment begins."[3] Hunting wild animals constitutes a fundamental connection between people and the land; as much as it was about elk, the confrontation in Jackson Hole in the summer of 1895 was about who was allowed to use the land, and how.

But if the whites and Bannocks were connected to the land through elk, what are the implications of that connection, and the struggle over it, for American history? If the conflict of 1895 looks so much like an incident out of an earlier West, then perhaps it should not surprise us that the first attempt to make historical sense of it was dismissive. The Supreme Court, deciding to strip the Bannocks of their hunting rights the following year, in 1896 relegated the Bannock experience to the status of denouement in the epic of American history. The court noted that much had changed since the Bannock peace treaty of 1868. At that time, "the progress of the white settlements westward had hardly, except in a very scattered way, reached the confines of the place selected for the Indian reservation." Even in 1868, the court said, "the march of advancing civilization foreshadowed the fact that the wilderness which lay on all sides" of the reservation "was destined to be occupied and settled by the white man, hence interfering with the hitherto untrammeled right of occupancy of the Indian." Now, in the 1890s, that settlement had come to pass. The court decided against the Indians on the grounds that wildlife in the six-year-old state of Wyoming was now the property of the state and its people. New settlement had created a new civil order. With the passing of the old order the Bannock hunting grounds were legally closed.[4]

However the court's decision measures up as a rendering of justice, as a historical interpretation it bears some remarkable similarities to the most famous interpretation of America's western past. Not two years before the tragedy at Fall River, Frederick Jackson Turner delivered his essay "The Significance of the Frontier in American History" in Chicago, at once demonstrating the historical significance of the western frontier and bidding it farewell. The movement of white settlers into Jackson Hole, which to the Supreme Court was the most significant event in the region's recent history, was part of westward expansion that Turner saw as the seminal process in American history. "The existence of an area of free land, its continuous recession, and the advance of American settlement westward, explain American development."[5] To Turner, the westering line of Euro-American settlement was the setting for universal processes of social evolution. On the frontier, American settlement moved through stages of development, from hunting, to stock raising, to farming, and finally to industrialization. It was an interpretation that cogently expressed ideas popular at the time, which helps to explain the similarity between Turner's writing and the Supreme Court opinion of 1896. There is no evidence to suggest that any of the Supreme Court justices knew about Turner's essay, but they obviously shared Turner's assumptions about the superiority of white civilization and the inevitability of white dominance in the West.

Partly because of those ideas, Turner's vision of the past has long been discredited, and in any case it does little to explain what happened at Jackson Hole in 1895, five years after the end of Turner's frontier. It seems hardly enlightening to conclude that Fall River on that July morning was, as Turner wrote of the frontier, "the meeting point between savagery and civilization." And his stages of social evolution seem curiously inverted in this story, the march of "civilization" regressing, with stockmen and farmers killing Indians to secure better hunting.

But if we step back from the Turnerian notions of social evolution and look again at the Supreme Court decision of 1896, we may find some materials for a new, more useful interpretation of what happened at Fall River. Ignoring the racial overtones of their logic, we can see that Turner and the justices shared a focus on Euro-American settlement of the West and on its consequences for how people lived on the land and apportioned its bounty. Such concerns are not unique to the nineteenth century. In our own era, the impact of growing populations on resource allocation is frequently a subject of debate in political and academic circles. One of the most provocative scholarly contributions on the subject is "The Tragedy of the Commons," an influential essay on natural resource policy and philosophy by the biologist and human ecologist Garrett Hardin. Written in 1968, Hardin's work drew a dark picture of a growing human population with open access to fragile and rapidly diminishing world resources. Hardin did not cite Turner, but when read alongside Turner's, his work begins to look oddly Turnerian at several crucial points. Taken together, their works begin to suggest how we might place the struggle over elk at Jackson Hole in a historical narrative that connects the Bannock tragedy not only to the nineteenth-century West of Indian wars and the open range, but also to the twentieth-century West of fences and the dominant power of federal agencies.

Hardin began by inviting the reader to "picture a pasture open to all," and went on to explain why a social system founded on free access to resources was not sustainable. His model had a compelling narrative: although tribal wars, poaching, and disease might protect the hypothetical commons from over-stocking for some time, ultimately, "comes the day of reckoning . . . the day when the long-desired goal of social stability becomes a reality. At this point, the inherent logic of the commons remorselessly generates tragedy." As a "rational being," each herdsman seeks "to maximize his gain." Sooner or later, each realizes that he acquires all the proceeds from the sale of a cow, but that the damage his cows do to the commons is shared with the rest of the village. The only "sensible" course for him to pursue is to put as many animals on the commons as possible, for the individual's gain far outweighs the individual's loss. Overstocked, the commons is ultimately overgrazed. Thus, Hardin observed, "Freedom in a commons brings ruin to all."[6]

Like other secular prophecies, Hardin's vision was founded on a historical interpretation. "As the human population has increased," he argued, "the commons has had to be abandoned in one aspect after another." Human history became the story of a global commons increasingly hedged as more people settled on it. Beginning with landscapes that produced food—farm land, pastures, hunting and fishing areas—humans had given up on common property and put their faith in private property or regulations that functioned like private property. The process continued in the late twentieth century. In lands and waters used for sewage disposal and in air where atmospheric pollutants were dumped, people steadily abandoned the commons as settlement increased. The surging human population of the world led Hardin to a dismal and seemingly inevitable conclusion: another encroachment on the commons was at hand. The freedom to bear children who subsisted on the global commons was as disastrous as the freedom to graze cattle on the village common. It was time, said Hardin, for humanity to abandon "the commons in breeding." Free access to critical resources must give way to restraint, "mutual coercion mutually agreed upon."[7] Hardin's historical vision was one in which expanding, free societies must become less free to ensure biological survival.

An apocalyptic warning in a dark American year, Hardin's 1968 essay was tremendously influential. Since its publication it has spawned a vast inter-disciplinary literature, most of which tests Hardin's thesis or evaluates its im-plications for resource policy or philosophy.[8] Some have focused on its useful-ness as a historical model, recognizing that, at heart, "The Tragedy of the Commons" is historical narrative—or fable.[9]

Indeed, it is striking how much the dynamics of Hardin's "tragedy" parallel the processes at the center of the classic "frontier" thesis of western history. Hardin defines free access to scarce resources as the defining characteristic of human history; the frontier historians, led by Turner, established free access to land as the defining condition of American history. In a sense, Turner's "free land" is the counterpart to Hardin's commons, "a pasture open to all."[10]

There are deeper parallels as well. To Turner and Hardin, the story of civilization could be read in how abundant resources, open to everybody, were absorbed into human economy. In essence, Turner's frontier processes were stages in the enclosure of abundant, open-access resources. It became the prin-cipal contention of the frontier school of history that the changes required of American society in "winning a wilderness," or enclosing the western com-mons, were the principal formative influence on American institutions, culture, and life; Hardin's essay postulated that removing resources from the commons and making them private property, or something very much like it, was the defining process of human history.[11]

A comparison of their works should not understate their differences: Har-din argued that more of the commons needed closing in order to avoid ecologi-cal devastation; Turner's essay was a paean to the commons upon its closure, and he spoke more to a contemporary belief in the wonders of progress than to wor-ries about environmental decay. Still, there were larger similarities. In both essays, expanding settlement on the land brought about the end of free access to resources; both authors viewed the closure of the commons or frontier as inevi-table; both believed that American society became less free when it happened.

But Hardin's essay, like Turner's, is ultimately not very useful for under-standing history in the American West. So similar are the failings of the two that in many cases criticisms of one thesis apply equally well to the other: in relying completely on Anglo-American notions of "civilization" and "advancement" to make its point, Turner's argument was decidedly ethnocentric; in failing to consider non-Western notions of property and community, so was Hardin's. Turner's "stages" in social evolution on the frontier were rigid and practically useless for illuminating real historical events, as were Hardin's ideas about "abandoning" various "commonses" in farming, grazing, fishing, hunting, and waste disposal. Perhaps the greatest faults of the two essays were their respective failures of definition: where Turner did not define "frontier" in a fashion that could be rigorously applied and tested, Hardin's notion of "commons" is a vast oversimplification, which not only fails to consider the ecological context of resources but also relies on a vision of "open-access" that is historically false. Rarely has any society allowed all members completely free access to any resource. As Turner's critics often point out, not even land in the American West was free.[12]

But for all their flaws, these two essays point up the centrality of free access to abundant resources as an organizing concept in history, particularly in the American West. If Turner's claims were overdrawn, others have argued convincingly that the frontier as a synonym for abundance does much to explain fundamental features of American history.[13] This approach leaves some important questions unresolved. By themselves, the lands of the West did not signify abundance. For that to happen, societies, especially the Euro-American society which provided most of the immigrants to the region, had to find a way of transforming them into usable resources. California's Colorado Desert was not a landscape of abundance. Irrigation turned it into the Imperial Valley, which was. Similarly, the Great Plains was not a viable home for many Indian peoples until the horse made its distances easier to cover and its buffalo more accessible. By the eighteenth century, even the Bannocks could ride to the distant Plains for the fall buffalo hunt.[14] The story of abundance and the frontier must encompass cultural developments that make the land produce for the society in question, that make it possible to wrest "the good life" from the land.

A study that acknowledged the connection between changing cultures, changing ecology, and the commons was Arthur McEvoy's history of the California fisheries, *The Fisherman's Problem* (1986). Although McEvoy does not directly address Turner's thesis, his work calls into question simple notions of "opening" and "closing" the commons. In fact, McEvoy's environmental history of the fisheries portrays a state- and federally regulated commons emerging out of various earlier and more localized and ethnically bounded commons regimes. From Indians' complicated and culturally regulated fishing of the annual salmon runs, to various immigrant uses of the riverine and ocean fisheries and clam beds, different local groups maintained distinctive relations with the fisheries. Ultimately, the rise of state and federal power over the fisheries signals the creation of a larger, centralized commons, one that government authorities regulate in an attempt to ensure relative abundance of fish for all users.[15]

McEvoy does not relate his thesis to land resources, and he does not explore dimensions of local resistance to the centralized commons. But his approach suggests how we might draw historical connections between notions of abundance, the role of the commons in western history, and what happened at Fall River. In the American West, the diversity of human societies contributed to a panoply of notions of abundance. Different peoples lived on the land in very different ways. The tremendous variety of societies and cultures in the West—Hispanos, Chinese, Anglos, distinctive Indian polities, and others—suggests that the West was not one great commons, but numerous local commons regimes, in which local communities defined their own interactions with their natural environments.[16] Elk did not mean the same thing to the Bannocks as they did to the white settlers of Jackson Hole; wild animals did not mean the same thing to a Sioux winter camp as they did to a polyglot mining town or to a Mormon settlement. Questions of "proper" interaction with the local ecosystem—issues of who hunted, gathered wood, or used the water—were framed within local cultures with unique characteristics, and specific ways of regulating social activity on the land.

Unlike Hardin's commons, which fails to consider social and cultural restraints in common use rights, the local commons was not "open to all."[17] Often, at its perimeters, ethnic boundaries demarcated the local commons. Any Cheyenne or Sioux knew the ramifications of encountering a Crow or a Pawnee in the local buffalo commons. The Bannocks fought other ethnic groups who crossed into hunting areas they claimed as their own.[18] One of the dominant features of western history is the struggle between local polities and outsiders (nonlocals) for exclusive control over land and the critical resources on it.[19]

Local systems for allotting local resources were not utopian arrangements: internal boundaries divided the local commons among local people. Within communities, there were often class, gender, and generational divisions about who killed deer, who skinned them, and who got the best cuts of meat; there were similar cultural systems for determining who got the best and most land, who got the best wood lots, who was allowed to graze their cattle and where and for how long. Internal divisions on the Bannock commons were apparent in 1895, when observers reported that women dressed the elk carcasses and treated the hides.[20] Presumably, the men did the hunting. Such cultural traditions were in part designed to ensure local abundance, to meet local definitions of the good life.

The commons was a cultural construct, a way of interpreting and living on the land. But inevitably the land also shaped the culture. Local understanding of local ecology was usually incomplete; local control over the land was tenuous at best.[21] Local ecosystems therefore became historically active, responding in complex and unpredictable ways to human activity and other developments. The growth of Indian horse herds in the nineteenth century depleted range land and the numbers of buffalo the range supported.[22] Changes in vegetation stands had potentially profound consequences for people who depended on plants for food and wood. In the Jackson Hole area, Bannock fires, both intentional and otherwise, shaped the landscape by burning off mature brush and trees, creating clearings where new shoots could grow.[23] The new growth provided forage for elk.[24] The local commons was a social system that shaped the land, but local ecosystems also shaped the commons by defining the range of possibilities for human action there.[25] Local ecosystem and local culture continually shaped each other and acted upon each other.

The local commons characterized the nineteenth-century West, but the twentieth century was dramatically different. From the patchwork of local commons regimes that dominated the western landscape, we have arrived at a system of natural resource administration in which centralized powers play a large often dominant role. In timber, water, grazing areas, and game, state and federal governments have assumed powers over human relations with the land that were once within the purview of local communities. A measure of this transformation is the fact that much of what was the Bannock elk commons is now Grand Teton National Park and the National Elk Refuge.

How did this happen? As other scholars have observed, the federal government was a significant presence in the lives of nineteenth-century westerners, who endured and encouraged Washington's centrality in Indian policy, land allotment, and railroad development. But during the twentieth century, westerners witnessed a surge of federal power over forests, water, public lands, and wildlife. During the twentieth century conservationists reordered local-national relationships, wresting natural resources from local control into the hands of scientists and technicians, who funneled them into the rapidly expanding market economy.[26] Although it was partly an effort to ensure sustainability of resources, this passage of the local commons occurred not because of some "tragic" (as Hardin would have it) dynamic inherent to common property. It was the result of a process of market expansion in the West stretching back several hundred years. Indeed, the single greatest agent in transforming the local commons was trade, which created a network of exchanges among local commons regimes. Whether it was the exchange of beaver pelts for wampum in colonial New England or the modern-day sale of timber from national forests, markets were the principal motor propelling the local commons into state and national management regimes. The market is the great unspoken presence in Hardin's version of the commons, the prerequisite influence that he mistakenly calls "social stability." It is the market for cattle, not peace, that offers the "rational herdsman" a way to turn his cattle to cash, thereby beginning the supposedly tragic sequence. Likewise, in Turner's essay, the story of the frontier is the story of market integration of the countryside, in which successive waves of entrepreneurs market furs, cattle, minerals, and crops all taken from the bounty of the Great West.[27]

The explosion of market networks with the advent of the railroad brought fears of environmental devastation and, in the late 1800s, official attempts to regulate resource use in new ways. Through conservation, local timber became property of the U.S. Forest Service, local water was diverted into federal irrigation projects, local land came under the aegis of the Bureau of Land Management, and state and federal agents stood between hunters and local game. More than any other time in the region's past, the twentieth century was when what was local became national.[28]

This, then, is the story at the heart of America's western past: the local commons giving way to the extra-local, the community surrendering authority in resource allocation to state or national agents. To its proponents, the ascendance of extra-local authority over resources was essential to democratizing them. In the words of turn-of-the-century conservationist Gifford Pinchot, "the natural resources of the Nation" now belonged "to all the people."[29] Local game and other natural resources now became not so much local property as the domain of an abstract American "public"; the wildlife of the local commons became a kind of "public goods."

Conservationists almost always defined this public in vague terms. Local communities were comprised of individual people, living in bounded settlements at specific locations. In contrast, the public was amorphous, comprised of nameless masses and bounded only by national borders. No local claim could transcend the public welfare; no local commons could supercede the public good to claim the public's game. By abstracting their constituency in this way, conservationists had a powerful foil to spatially bounded local prerogatives. Where natural resources were concerned, proximity conveyed less privilege than ever before.

In this sense, conservation implied something more than making resources like wildlife into a form of state property. Deer and elk were not just a national holding akin to a post office, a military base, or an interstate highway. For its champions, conservation offered a utopian vision, a national version of local community use rights to the land, a "national commons." To be sure, the ideal of a national commons was seldom if ever realized. Scientific management often removed user control over resources and imposed bureaucratic design with little reference to public wishes. It could result in near-capture of resources by industry interest groups and even virtual privatization of use rights.[30]

Scientific management also resulted in a fundamental abstraction of resources from place. The local commons was tied to a specific place, a local area where local use rights prevailed. Conservationists often divorced rights of access from place to the extent that the owners of private land on which public resources were found were no more entitled to them than the public at large. This was especially true of wildlife, which was public property managed by the state, even on private land. In this sense, the commons underwent perhaps its most significant transformation. The commons was no longer a piece of land but a pool of wildlife or other resources in the public hands. To look across private land and see the public's game grazing there was to realize that one did not live on the commons, but beside it, or with it.

By defining animals as public goods, conservationists deflected local charges that poor people were being shunted from the hunting grounds to make them a playground where the rich could recreate. Whether or not a genuine national commons was ever a possibility, conservationists could defend the increased authority of extra-local agents over local game as the necessary intervention of the state in the management of public goods. Conservationist strategy on this point was strongly influenced by popular understandings of the European experience in game protection. Throughout western Europe and particularly in England, local communities had seen their historic common

areas enclosed by wealthy landlords, who also claimed the once-common wildlife as their private property. The state had become an ally in this effort, and the
subsequent centuries-long, low-level war between poachers and estate game
wardens was accompanied by left-wing allegations of class bias and elitism in
state legal structures.[31]

In America, the reification of wild animals as public goods immunized the
state against such charges. Rather than privatizing the game, American wild
animals were held in common by the public, for whom game law authorities
were merely agents. To label wildlife merely another form of state property
obscures this subtle blending of common property and state authority that
conservation embodied. In communities across America, the collisions between
local rights and public claims gave rise to a great deal of shouting, and considerable shooting. Conservation is such a broad subject that wood, water, or one of
many other resources could have been the subject for this study. Yet there are
compelling reasons for examining wildlife. Unlike timber or land, animals
move from place to place, often unpredictably. Bundles of use rights attached to
game might accompany them or shift radically as they cross boundaries between states, between public and private land, between ocean and river. Wildlife may seem a peripheral subject insofar as it is of marginal economic importance to most of America today, but, as James Scott has observed, when the core
modes of production are dominated by one class, resistance to that class may be
found in the margins of the economy as well as in the centers.[32]

Perhaps because of the deep and conflicting beliefs that many people
attach to wild animals, disputes over hunting tend to be particularly explosive,
underscoring in angry hue the kinds of conflict inherent to conservation generally. A brief survey of the history of wildlife conservation at the turn of the
century reveals the depth and complexity of conservation struggles and intro-
duces some of the broader issues in the case studies that follow.

As game populations declined in the late nineteenth century, more hunters
turned to the remaining animals to meet the demand: local subsistence and
market hunters in America faced off against elite, urban "sportsmen" who
demanded state and federal regulation of hunting in accordance with their own
ideology of the hunt. Derived from aristocratic European traditions of "the
chase," the so-called sportsman's code was, in fact, a loose set of often conflicting ideals about masculine behavior in the hunting grounds. To some, hunting
deer with hounds was great sport; to others, stalking deer provided the only true
test of manly skill.[33] But for all their differences, recreational hunters came to

agree on certain general principles. Frowning on the killing of female and young deer as unmanly, market hunting as greedy, and year-round hunting as wasteful, sportsmen conservationists secured laws at the state and federal levels to enforce their own notions of proper hunting against the "game hogs" and "pot hunters" of local America.

Subsequent confrontations between local hunters and extra-local conservationists were nothing less than struggles over the place of human society in the natural world. In the nineteenth century, middle- and upper-class American men were preoccupied with the problem of defining masculinity in a world that seemed increasingly feminized. For many men, hunting became a symbol of masculine strength. How one hunted and what one killed came to define what kind of man one was. To a degree, elite development of these masculine ideals facilitated the emergence of conservation. In positing idealized masculine behaviors as "proper," elite recreational hunters began to set standards of behavior for other hunters, cleansing the fields of all who did not abide by the sportsmen's ethic. The backbone of the conservationist cause, recreational hunters lobbied lawmakers to pass game laws that effectively wrote their own sporting ideals into legal statutes.[34]

Because hunters were almost universally men, disputes over what constituted "proper" hunting were clashes of rival masculine ideals. Killing female deer or songbirds might have been considered unmanly, even cowardly, in sporting circles, but for many locals it was part of life. For Pueblo Indian men in New Mexico, bringing home wagons loaded with hundreds of deer—bucks, does, and fawns—was a masculine achievement, ensuring as it did the well-being of the community; for Italian immigrant men, a single shotgun blast could bring down a brace of sparrows or song birds, a frugal way of providing meat for the family table. Conservation bestowed the mantle of state power on sportsmen, enshrining their ideas of masculinity and outlawing behaviors deemed proper and manly among Pueblos, Italian immigrants, and others. In the collisions between local hunters and state authority, at stake was not only the political power of local people, but the behaviors, beliefs, and activities associated with being a man.

Fall River was only one such collision in a much larger pattern, often unnoticed by American historians.[35] Such events were not confined to the West: Florida's Guy Bradley became one of the first game wardens to die at the hands of a poacher, in 1905, and Pennsylvania lost four game wardens in shootouts with poachers in 1906.[36] In the pages to follow, I will compare eastern game conflicts to western ones and point up the important differences between these struggles. For now, I will merely suggest that western struggles were generally more "federalized"—in that they often involved conflicts between local hunters and national authorities—whereas eastern struggles involved people closer to home. In its national parks and national forests and their wild animal populations, the West became the primary testing ground for conservationist ideas of resource management, ideas that involved close ties among local landscapes, the federal government, and markets connecting the local commons to the urban centers of Europe and the American East. Conservationist ideas, innocuous in theory, were often executed in less benign ways. Officials enforcing game laws opened fire on Indian hunters, with deadly result, in Wyoming in 1903 and in Montana in 1908. Game authorities openly encouraged New Mexico ranchers to kill Indian hunters in 1905. Local whites at Yellowstone National Park had angry confrontations with army guards over hunting rights in this customary hunting area in the 1890s; local whites and Blackfeet Indians have contested federal authority in their traditional hunting grounds at Glacier National Park in northwestern Montana since the early twentieth century.[37]

In myriad ways, then, the story of the fateful encounter at Fall River repeated itself in different locales, at various times, all over the West. Where one community fought for access to game, another sought rights to timber, water, or land. The sheer diversity of local interests militated against their unification in a single cause of resistance to extra-local authority; the question of who gained and who lost with the advent of conservation meant that local communities were often deeply divided over resource issues. But, broadly speaking, the numerous collisions between local communities and extra-local authorities suggest a unified narrative connecting the nineteenth-century "frontier" West to the West of the twentieth century. Whether Indians facing annihilation by federal cavalry, cattle barons whose preeminence gave way to the Bureau of Land Management, or loggers in the Pacific Northwest who today face off against advocates of the Endangered Species Act, much of western experience can be explored with a simple question: what happened when the local commons became the state and federal commons?

Answering that question as it relates to wildlife is the project of this book. Before moving on to case studies in Pennsylvania, New Mexico, and Montana, let me a offer a definition of "local commons" and three premises.

Originating in the Latin *locus,* or place, the word *local* presents us with almost as many problems as opportunities. One of the word's attractions is its pervasive use in everyday life: people inevitably identify themselves in part as "local," expressing their relation to the place they call home. To the extent that this book is concerned with the politics of place, I find the use of "local" and "extra-local" to be suggestive of relations to particular settings.

But by itself, the word *local* masks varying degrees of rootedness. Some people are more "local" than others by virtue of family tenure in a specific location. And local identities often shift in the stories in the following pages. Locals in one area move to a new area, where they arrive as extra-locals. Taking up residence, some are soon claiming local status and demanding extra-local (often federal) assistance for the resolution of local problems that their own arrival has precipitated. Other "new locals" respond to the hostility of prior locals, who inevitably consider the newcomers extra-local. Rather than ontological states of being, the terms *local* and *extra-local* are at opposite ends of a scale of relation to place and community, along which a person can move back and forth many times during a lifetime.

A grasp of how local people have ordered their relations to the land is possible only if we construe those relations in a somewhat different fashion than Garrett Hardin did. There are alternative models: in the past decade, common property revisionists have redefined the commons as a resource managed by an identifiable community of users who exclude outsiders.[38] In a somewhat looser construction, I use *local commons* to denote local systems for apportioning resources that are not privately owned. Locals often define common property in informal ways; boundaries around resources and local regulations on resource use—especially social constraints—are often obscure or even invisible to outsiders. Compared to a bureaucratic resource management regime, a local commons may have few regulations. It may even tolerate outsiders, and in extreme cases the difference between a local commons and an open-access regime can become blurry.

More often, though, local communities prop up boundaries around resources in some fashion or another, facilitating locals' use of resources and impeding access by outsiders. These boundaries do not necessarily serve as a coherent management regime, yet they are systems of resource use, embedded in the local landscape by local people, serving to exclude extra-local people. Indeed, it is fair to say that locals in this book often view their own access to resources as the single greatest priority in allocating them, a position that frequently puts them at odds with scientists, bureaucrats, and especially conservationists.

The flexibility of the local commons is apparent in the wide array of strategies available for defending it. At times, locals chose aggressive confrontation, but more often locals have found their use of the commons under attack from people who were vastly more powerful than they. In such circumstances, the boundaries of the local commons were inscribed through nonconfrontational means: avoidance of outsiders, utilization of resources in which dominant actors had no interest, reticence, quiet resentment, misinformation. These and countless other informal means of boundary making did much to determine the shape of the local commons as communities resisted its incorporation into markets and government management regimes.

Three initial premises also help to clarify my interpretation. First, conservation was not imposed on all locals as brutally as Constable Manning's posse imposed it on the Bannocks. Usually, local conflicts over resources preceded the installation of federal regulations, and often, these conflicts were themselves the consequence of rapid market penetration of the countryside. The expansion of the railroad and attendant commodification of the landscape during the nineteenth century wrought such changes in economic and social relations that communities often became divided over how best to interact with the land. Where one local faction sought to avoid some implications of market expansion, other factions, usually elites, frequently requested outside regulation of resources as a means of ordering resource use toward their own ends. Not surprisingly, local society often fractured along class, race, and ethnic lines. When local whites at Fall River requested federal help in pushing Indians from the hunting grounds, they engaged in a common western strategy. Today, the heated rhetoric of local control that pervades the rural West often obscures the fact that a federal presence in resource administration emerged as part of local custom and culture, along with ranching and other forms of resource extraction.

Second, local resistance to the process was so tenacious that a version of the local commons sometimes found its way into the conservationist resource regime. Extra-local authorities could respond to local demands for customary prerogatives by turning a blind eye to some local law breaking, partly in resignation but partly to ensure some local support for enforcement of other legal proscriptions.

Third, as much as it was a dialogue between local people and extra-local interests, the transmogrification of the local commons encompassed an interchange between people and land. Ecosystems were dynamic, and neither locals nor extra-locals could control them. Ranchers in Jackson Hole realized this all too well only fifteen years after the massacre at Fall River. By 1909, many were complaining that thousands of starving elk would devour their hay fields. They demanded federal action, and the federal government responded with the creation of the National Elk Refuge—lands set aside to grow food for the elk and keep them away from ranchers' property.[39] The dynamism of wild animals, plant communities, and forests—their tendency to surge in population over a relatively brief period and to level off or decline a short time later—ensures that the land is never really "settled." Simply put, the local natural environment frequently acts to "unsettle" the land, which changes as we live on it. Even today, drought on the plains, rust blight in the forests, and irruptions in elk populations can send locals and extra-locals alike scrambling for new systems of allocation to suit the changing land.[40]

Western history has long been troubled by the seeming disjuncture between the history of the region before 1890, the "frontier" period of Indian wars and cattle drives, and the history of the twentieth-century West, when the region became characterized by the dominance of federal agencies, urban settlement, and fences. A number of recent works have sought to unite the disparate stories of these two Wests. Patricia Nelson Limerick has argued that processes of conquest that characterized the West during the nineteenth century continue today; Richard White has concluded that as early as the nineteenth century the region's reliance on federal largesse allowed the national government to develop in new ways, extending its reach and prefiguring the rise of federal power across America in the twentieth century; William Cronon has explored the transition of the West from nineteenth-century frontier to twentieth-century region by exploring the growth of metropolitan centers and their transformations of western hinterlands.[41]

Following the local commons on its journey into public goods—a national commons—allows us to connect the West's nineteenth- and twentieth-century histories in a continuous narrative that speaks to many of the same issues as these authors, while affording some additional advantages. Limerick's notion of conquest, while it has much merit, is often too vague to explain the kinds of subtle dialogue between peoples and dominant powers that took place on the commons. The growth of conservation at times contained elements of conquest, but more often locals retained too much power over the process for us to consider them "conquered." Exploring the movement of the local into the national suggests connections to White's processes of federalization and Cronon's analysis of market expansion. But at the same time it focuses on local experience, on the diverse and often conflicting ways in which local communities responded to the changes in daily living brought on by the federal- and market-driven transformation of the West.

Attention to issues of the "frontier" has obscured the subtlety of the commons and its history in the West. In seeking to find an "end" to frontier settlement at the beginning of the twentieth century, we obscure the connection between the local commons, which characterized the West of the nineteenth century, and the national commons, which dominates the West today and has been an essential part of it for the past hundred years. The continuing dialogue between local community and national authority—an exchange involving not only words, but such gestures as poaching, illegal grazing, and timber trespass—has given rise to a narrative tension in the history of the American West. The contest between the local and the national is more than figurative. It is a literal force in regional politics and in day-to-day living.[42]

The movement of the commons from local to national is comprehensible in retrospect. To those who lived through it, it could bring rewards, as it did to the hunting guides and dude ranchers of Jackson Hole. But for others, the experience could be bewildering and terrifying. For Ben Senowin and his companions, wildlife had been part of a distinctive, familiar local commons. The intrusion of outside authority into their relations with the natural world was a shattering moment, nowhere communicated better than in Senowin's affidavit before the summary court officer in September 1895. According to the court officer, Senowin testified that

> neither he nor any of his people were told why or by what authority they were assaulted; that he is not aware that either he or any of his party had committed any offense against the laws of any State or the United States; or that he or any of his party ever attempted or offered any violence, or had made any threats against the life or property of any white man; that the white man never gave him or his party any hearing, or asked him or his party any questions through an interpreter or otherwise; that neither he or any of his party were ever called upon to answer or plead in any court of justice or make answer to any charge whatsoever.[43]

Ben Senowin and his hunting party were people of the local commons. They were witnesses to the process that would make America's wildlife what it is today: public property, managed by state and federal authorities. Appreciating the story of the Bannocks brings us closer to understanding what it meant to move from the local to the national, to knowing the human experience behind the process called conservation. Although the next chapter considers events that transpired more than ten years after the Bannock encounter with the white posse, and a thousand miles east of Wyoming, it will take us toward a fuller apprehension of what happened at Fall River that summer day.

1
The Killing of Seely Houk

On April 24, 1906, a Pennsylvania Railroad engineer on a passing train spotted a body in the Mahoning River in Lawrence County, western Pennsylvania. He dropped a note to that effect at the next station, Hillsville, a small limestone quarrying settlement about ten miles west of New Castle, near the Ohio state line. The telegraph operator retrieved the note and sent a line crew to investigate. Shortly thereafter, the body of L. Seely Houk, the deputy game protector of Lawrence County who had been missing for more than a month, was pulled from the Mahoning with at least two shotgun blasts in his head and torso. Whoever killed Houk did not want his body found: it had been weighed down with large stones, becoming visible only when the water level fell as the spring runoff subsided.[1]

The discovery of Houk's body began one of the most bizarre series of events in the annals of wildlife conservation. Suspicion of murder immediately fell on the residents of Hillsville, most of whom were Italian immigrant quarry workers. Pinkerton detectives hired by the Game Commission soon infiltrated what they called the Hillsville Black Hand, a secret society comprised of Italians known to indulge in criminal activity, especially extortion. One Pinkerton Agency spy lived in Hillsville for a year and a half in order to become a high-ranking society member, while secretly sending important tips to legal authorities. Ultimately—after another murder, dozens of arrests, and a temporary occupation of the town by state police—one local Italian man, the former leader of the so-called Black Hand in Hillsville, went to the gallows for the murder of Seely Houk. Immigrants continued to feel the aftershocks for a long time. In 1909, the Houk murder still fresh in their minds, the state legislature outlawed possession of a gun by any noncitizen immigrant. The Alien Gun Law remained in force for more than forty years.

Around the offices of the Game Commission, the story of the Houk murder became official lore, testament to the tenacity and toughness of early conservationist stalwarts, who faced down even Italian gangsters to save wildlife.[2] But, as useful as the legend was for authorities, to a historian it cries out for explanation. What brought about this bloody turn of events? "Black Hand" organizations are commonly understood as early manifestations of Italian-American mafia, and they are usually associated with crime in the Little Italies of America's cities. Knowledge of such organizations is incomplete at best: sources are few, speculations rife. Authorities at the turn of the century often labeled any Italian immigrant crime as Black Hand activity. But was there something we can identify as a mafia operating in this remote immigrant settlement in 1906? It was not until Prohibition that mafia families began to move into the front ranks of criminal syndicates. What social purposes could a mafia organization serve, what gaps in civic structure could it exploit, over a decade before Prohibition and miles from the larger cities where mafia firms would one day flourish? And when the showdown with state authorities finally came, why was it over rabbits, sparrows, and woodchucks, not boot-legging, prostitution, or extortion? Hillsville was not the only place where Italian immigrants and Pennsylvania game wardens met in violent confrontations. The Game Commission reported other murders, several in 1906 alone. Joseph Kalbfus, head of the Pennsylvania Game Commission, received three threatening letters from Black Hand correspondents in the early 1900s.[3] But the Hillsville killings were the only ones solved, and apparently the only ones from which any evidence survives.

The conflict between conservationists and Italians in Hillsville was partly a contest between extra-local authorities and local hunters. The wildlife habitat in Hillsville's environs consisted of private farms. The commons at issue was not the hunting grounds itself but the game on it—the rabbits, groundhogs, and birds to which the state and immigrants staked rival claims. But the struggle in Hillsville was more complicated than this. The confrontation was essentially triangular, among Italian immigrant workers, native-born farmers, and the state. Pennsylvania's rural English-speaking populace had long regulated their own interactions with wildlife, and even state laws were enforced solely by local constables and aldermen. The arrival of large numbers of immigrants in Hillsville had changed all this, as Italian hunters roamed farm fields, contributing to a sense—pervasive in some quarters—that the countryside was under siege by armed foreigners.

The landowners of Lawrence County responded by calling on the newly appointed state game protection officer, L. Seely Houk, who set about rigorously enforcing state game laws. Consequent hardships for immigrants motivated some to violence. The presence of game law officers in immigrant communities was an unprecedented intrusion of state power into local life, threatening the political culture of immigrant settlements in new and sometimes explosive ways.

As an eastern case study, circumstances surrounding the Houk murder suggest that eastern states experienced many of the same tensions over game laws that so fractured western society. Western conservation struggles differed radically from those in the East (see chapters 3–6), but we can learn a tremendous amount about the commons and the social dynamite of conservation by pursuing the mystery that so puzzled authorities in the spring of 1906. To wit, who killed Seely Houk?

The Killing

Like many other states, Pennsylvania began imposing new laws to protect wildlife during the late nineteenth century. By 1900, the state had imposed bag limits, set closed seasons on hunting, and outlawed the hunting of many species, especially songbirds.[4] For the most part, the legislature intended for these laws to be enforced by local police authorities: the first state game wardens were not appointed until 1895, and even then there were only ten of them. Seeking to broaden the state's enforcement apparatus, the Game Commission lobbied successfully for the creation of a new subsidiary class of wardens, called deputy game protectors. Beginning in 1903, each county could appoint a deputy game protector, bonded and possessing the powers of a regular warden. The principal difference between the two classes of wardens was their pay: deputy protectors received no salary but were entitled to half the fines assessed any violators they caught.[5]

Seely Houk's appointment as deputy game protector of Lawrence County in 1903 brought state authority into the hunting fields around Hillsville for the first time, a development that some local hunters soon resented. After Houk's murder, detectives interviewed local aldermen—who served as judges in poaching cases—to see who might have wanted to kill Houk. They found many potential suspects, all of them Italian. Italian immigrant hunters had been frequent targets of Seely Houk's patrols, and his characteristic aggression in performing his duties often made his arrests confrontational.[6] Alderman Charles Haus adjudicated many of the poaching cases Houk brought to court, and, as he told the Pinkertons, "Houk was a man who was absolutely without fear and, although it was generally understood that his life had been threatened by many Italians, it had never interfered with his work as an officer." Indeed, Houk preferred brinkmanship to compromise. "Houk was a man who was likely to give anyone who threatened him every opportunity to carry out their threat, provided they were quicker in drawing their gun than he was." This would not have been easy, for "Houk was noted for his ability to draw his gun quickly," and in "numerous instances" he had confiscated guns from Italians and brought them to the alderman's office.[7]

Houk's run-ins with Italian poachers were not isolated incidents. The game warden was part of a growing state apparatus intended to seal farm country not just against Italian poachers, but against whole classes of rural and urban hunters. The origins of this campaign lay with rural landowners. In the late 1800s, as railroads proliferated across the countryside, many Pennsylvania farmers began to feel threatened by visitors from once-distant cities and suburbs. Complaints about beggars, thieves, hoboes, and poachers inundated local constables, who found themselves unable to prosecute many cases because alleged criminals frequently fled beyond local jurisdiction on the train. In response to the increasingly cosmopolitan, delocalized nature of rural crime, the legislature created the Pennsylvania State Police, a mounted force with powers to enforce state laws in any county or city.

From their earliest days, much of the state police's energy went to enforcing game laws, for landowner complaints about poachers were particularly vehement.[8] Within farm communities hunting was an established tradition, but many rural landowners regarded city and out-of-state hunters as armed trespassers, at best a nuisance and at worst life-threatening. And the railroad brought such hunters to the countryside in droves. At the turn of the century, many farmers complained that their land "both in and out of season" was "run over by irresponsible hunters from adjoining states, who tore down fences, shot poultry, crippled stock, started fires and committed other depredations, then quietly disappeared into their own territory safe from pursuit."[9]

Given this context, it is not surprising that the Game Commission—seeking to protect the game better in farm country—worked with the state legislature to curb the predations of extra-local hunters. In 1901 the state increased legal penalties for armed trespassing and for the first time required hunters from out of state to buy a license before hunting in Pennsylvania.[10]

Farmers were seeking better defense of private property, but in turn-of-the-century Pennsylvania, private property and common property complemented each other to such a degree that it is impossible to understand one without recognizing the role of the other. Indeed, at its heart this confrontation between farmers and hunters was a struggle between alternative common property regimes. The new hunting restrictions meant that the commons as a place was diminished, insofar as hunters who had used private land as a de facto common hunting area now faced greater penalties for doing so. At the same time, in propounding complex rules of access to deer herds and bird flocks—hunting licenses, closed seasons, and so forth—state authorities erected boundaries around deer and birds as a kind of common pool resource, regardless of where the resource actually was within the state. Almost all wildlife habitat was privately owned, but the wild animals were public goods: landowners had no more right to hunt them than did any other Pennsylvania hunter. Legally, the animals belonged to the public of Pennsylvania, on whose behalf the Game Commission managed and protected them. In this sense, they now constituted a kind of "state commons." There was much room for conflict between landowners and conservationists on the issue of the public's game, but for the time being, deputy game protectors and state policemen patrolled the "boundaries" of the state commons by prosecuting illegal incursions into private property.

Seely Houk's patrols were part of this development, but they also occurred against a backdrop of ethnic conflict. There was much antipathy between the Game Commission and Pennsylvania's Italian immigrants, and indeed between American conservationists and Italian immigrants in general. Early wildlife conservation in Pennsylvania—and throughout America—was characterized by strong nativism. Turn-of-the-century conservationist polemic often decried the "wasteful" and "greedy" hunting practices of certain ethnic and social groups, among them American Indians, blacks, and immigrants, especially Italians.[11] It is difficult to overstate the degree to which conservationists despised Italians for killing songbirds and insectivorous birds.[12] Part of the conservationist concern was aesthetic. Bird preservation rhetoric featured strong connections among songbirds, femininity, music, and healthy, well-rounded living. Many scholars and writers were convinced that music originated in ancient humanity's study of songbirds. The animals were, in this sense, the means through which culture germinated from nature. To be indifferent to the preservation of songbirds betrayed a callousness not just to the animals but to music and indeed all of "civilized" morality.[13] The point is perhaps best illustrated by Congressman John Lacey, one of America's foremost wildlife champions, who declared in a speech on the floor of the House of Representatives: "The man or the woman who does not love birds should be classed with the person who has no love for music—fit only for treasons, stratagems, and spoils. I would love to have a solo singer in every bush and a choir of birds in every tree top."[14]

Increasingly, naturalists attached economic as well as aesthetic values to that "choir of birds in every tree top." In Pennsylvania, as elsewhere, conservationists maintained that insectivorous birds were essential to the well-being of farm crops. They saw the dramatic destruction of the passenger pigeon and the carolina parakeet as signs that all birds were in danger of extinction from overhunting. Although conservationists embraced antipredator campaigns and the eradication of raptors like hawks and owls, they also agreed that the impending disappearance of insectivorous birds would bring ecological catastrophe. Farm crops would suffer, and so would other plants. Birds ate tree-destroying insects and helped distribute seeds; therefore, concluded one study for New York's state game authorities, "it can be clearly demonstrated that if we should lose our birds we should also lose our forests."[15] In an age when forest conservation was a byword for watershed protection and the maintenance of urban civilization, this was doomsday rhetoric. State authorities from Pennsylvania to New Mexico frequently drew similar pictures of environmental devastation in a nation without birds, where "this whole continent would in three years become uninhabitable by reason of the myriads of insects."[16]

Because Italian immigrants hunted birds, they were considered a principal threat to bird life and therefore an incipient cause of this potential apocalypse. In the minds of many, immigrants represented as much a threat to American nature, especially American birds, as they did to the social order of American cities. William T. Hornaday, the president of the New York Zoological Society and one of the most famous conservationists of the era, captured conservationists' fears of Italians in his widely read tract, *Our Vanishing Wildlife*: "Let every state and province in America look out sharply for the bird-killing foreigner; for sooner or later, he will surely attack your wild life. The Italians are spreading, spreading, spreading. If you are without them to-day, to-morrow they will be around you."[17]

The ethnic antagonism between English-speaking white Americans and Italian immigrants in the early 1900s extended far beyond the hunting grounds. The waning of the nineteenth century witnessed a surge in Italian immigration, especially to states in the Northeast, including Pennsylvania. Soon after 1880, large numbers of immigrants from central, southern, and eastern Europe began to arrive on American shores. Between 1900 and 1910 more than 2.1 million Italians arrived in America, and many went to Pennsylvania.[18]

Italian immigrants in Pennsylvania worked as coal miners, factory employees, and laborers.[19] Their arrival precipitated widespread ethnic tensions and a pervasive sense that American society was undergoing a profound transformation.[20] This social upheaval was pronounced in Lawrence County, where Italians arrived in large numbers to take up jobs in the steel and tin mills of New Castle. In 1890, the town's population stood at 11,000, with only a few Italian families. But by 1900, the population had increased over 150 percent, to almost 29,000, with Poles, Slovaks, and Italians comprising most of the increase. Most of the Italians came from southern Italy, and they took jobs on the railroads and in the steel mills and tin mills. West of New Castle, in Hillsville, they arrived to work the quarries, and by 1906 one observer estimated that the town's population of 1,500 included 900 Italians.[21]

Wage labor may have been the basis of survival for most immigrants, but in Hillsville and elsewhere—where a full day of blasting, hacking, and loading limestone brought $1.65—subsistence hunting was a supplement to wages and an important part of Italian immigrant life. Songbirds were a customary Italian delicacy. They were widely hunted in Italy and sold for food in public markets.[22] In Hillsville, Italian quarry workers often took guns to the woods after work. Particularly in the spring and summer, when the days were longer, they trespassed on surrounding farms, seeking out small game—groundhogs, rabbits, and songbirds—to provide meat for the table.[23]

These hunting trips were more than just a means of gathering food; they were also an escape from the noise and dust of the quarries. The lure of the hunt bonded neighbors and friends. In bed with a painful leg injury, one Pinkerton undercover agent tried to refuse an invitation to a hunting trip, complaining, "It is too hard for me to travel through the woods." The immigrants cajoled him into going anyway, because the party "did not need to go around much," because "they had a good dog with them . . . [and] we could sit down and the dog would bring the rabbits around." Ultimately, "We all went down toward Peanut Quarry angle line and around in the woods. We killed four rabbits."

Game meat like this was prized. "We went by Jas. Rich's house, and he wanted to cook the rabbits, but Joe Gallo . . . wanted his wife to cook them. Lowi Ritort said that he was going to take one to his girl's house and have a supper."[24] Elderly men in Hillsville today recall immigrants hunting groundhogs, rabbits, and "anything that flies," and their mothers cooking songbirds "just like they cook chickens."[25] Sometimes the meat was cooked with vegetables or tomato sauce, or added to a sauce for spaghetti.[26] Groundhog with potatoes made a filling and delicious stew.[27]

Important as it was to local Italians, this subsistence hunting clashed with many aspects of the conservationist program. Indeed, in outlawing Sunday hunting, the killing of all but a few species of designated game birds, and all methods of hunting other than with a gun, the conservationist program was designed in part to convert game from a year-round meat resource to a seasonal recreation.

Obviously, immigrants were not the only subsistence hunters in Pennsylvania at the turn of the century. The state's vast and rugged northern mountains were home to many people who hunted for food practically all year.[28] But Italians lived in segregated communities; they spoke another language, dressed differently, were easily identified, and were widely mistrusted. It did not take long for the new Game Commission to zero in on Italian hunters. As early as 1902, the commission reported that "the unnaturalized foreigner" was responsible for the great majority of game law violations, and that Italians were especially troublesome. The 1908 report of the Game Commission complained that "by far the greater number of cases of violation of our game laws reported to us during the past season, killing of game out of season, hunting on Sunday, killing song and insectivorous birds, is of wrongs done by the unnaturalized foreign born resident of this State, mostly Italians."[29]

And, where legislators had begun to push unlicensed, out-of-state hunters off the hunting grounds, they soon took action against the growing numbers of Italian hunters. In 1903—the same year Seely Houk became deputy game protector of Lawrence County—the commission secured passage of a new game law, which had ominous consequences for hunters in Hillsville and other immigrant communities. On its face, the law was a simple measure, requiring non-residents to purchase a $10 license before hunting (state residents did not need hunting licenses until 1912).

But the Non-Resident License Law, as it was called, contained a peculiar twist: it defined nonnaturalized immigrants—and therefore most of the people in Hillsville—as non-residents. For immigrants who did not buy a license, conviction could follow the simple act of carrying a gun "in the fields or in the forests or on the waters of this Commonwealth." Violations were punishable by a $25 fine. If unable to pay, the violator was to spend one day in jail for each dollar of the fine assessed.[30]

On top of the fines for violating the 1903 law, there were additional fines for violating other statutes. Thus, if a noncitizen hunter were caught with no license and three songbirds, the violator paid $25 for not having a license, and

$10 for each "nongame" bird. If caught hunting on a Sunday, a $25 fine was added to the others. Complaining about Italian poachers in 1905, Joseph Kalbfus, the secretary of the Game Commission, estimated that "an arrest of one of these people for violating the game laws seldom results in a penalty of less than $60 or $70 with costs, sometimes very much more than this amount."[31]

Wildlife had never been privately owned in Hillsville; local hunters had stalked farm fields for generations, asserting a kind of community ownership of game. Potentially, outside hunters may have also claimed a right to hunt the animals, but as rabbits and groundhogs inspired little interest beyond the community, local preeminence in the hunting grounds remained unchallenged, and wildlife served as a kind of local commons.

Essentially, immigrants had tried to reformulate the local commons in wildlife to suit their own cultural expectations, and in doing so they reshaped local patterns of hunting. To judge from the complaints of local farmers, the older system of hunting in the county had entailed a degree of personal connection between hunter and landowner, because the latter had to give permission for the former to hunt. Italians did not ask permission to shoot local game. And their taste in wildlife—songbirds and groundhogs—must have struck American-born locals as bizarre. The 1903 Non-Resident License Law was a state effort to extinguish this kind of cultural reorientation of local hunting practices and impose a state order in its place. The bounding of game as a "public good" entailed the exclusion of immigrants from the hunting public. By making it a crime for an immigrant to carry a gun without first paying $10 to the state, the law practically forbade most Italians from hunting at all. The law certainly had that effect in Hillsville.[32]

Conflict over the new game law aggravated an already tense situation in Hillsville. As in other parts of Pennsylvania, confrontations between immigrants and prior residents seemed to threaten the rural social order. The settlement was isolated in rolling country where sugar maple and beech trees grew thick on the hills sloping down to the Mahoning River.[33] The quarries that drew the Italian immigrants belonged to the Johnsons and the Duffs, local families who, like most other landowners in the area, were English-speaking farmers. Such people were perplexed and often frightened by the growing numbers of armed immigrants who stalked their fields, committing trespass at the same time they violated the game laws. Soon landowners were leveling charges of stock theft, vandalism, and intimidation against the newcomers.

The editor of the *New Castle News* commented on the social tension in the farm fields around Hillsville in a column published in 1907. The piece lauded the arrests of several Italian men in Hillsville, alleging that farmers in the area had been so terrorized by roving bands of armed Italians that "No person in the Hillsville district, either Italian or American, will give the slightest assistance to any officer desiring the prosecution of Italian offenders." The writer detailed incidents of intimidation in the Hillsville vicinity: a farmer who let an officer use his telephone to effect the arrest of an Italian found one of his cows shot dead the following morning, with a note in poorly written English saying, "This is for assisting the police"; the arrest of another Italian for trespassing brought dozens of Italians into the area, asking local residents if they knew the whereabouts of the farmer who called the authorities. Terrified, local farmers snuffed out their lanterns at night and posted armed sentries.[34] Such fears may have been overwrought, and the editorial may have been an example of the yellow journalism so popular during the period. But other evidence substantiates the deep fear and hostility that prevailed in the Hillsville region at the time of Houk's death. Children of Italian immigrants who lived in Hillsville in 1906 recall that some local immigrants committed stock theft and rustling in the area—"lots of that"—until sometime after the First World War.[35]

This was the social context of Houk's murder, and detectives on the case witnessed the simmering hostility between immigrant poachers and local landowners as it boiled over again. Only a few months after the discovery of Houk's body, three Italian men decided to go hunting. Taking along two shotguns, they walked onto the lands of farmer and quarry owner William "Squire" Duff. Duff, eighty years old, found them and ordered them to leave. Shortly afterward and while still on his property, they killed two small birds. Duff returned to order them away again. One of the hunters, Dominic Sianato, discharged his weapon in the old man's face.[36]

Local Italians considered the killing of Duff immoral and unacceptable. Unlike the Houk murder, which was veiled in secrecy, Hillsville residents talked openly about Duff's killing among themselves, enabling the Pinkerton undercover man—on the scene to investigate the Houk killing—to advise the authorities of the assailant's identity soon after the event.[37]

Duff died in August 1906, five months after Houk vanished while patrolling along the Mahoning. We cannot know what was going through Squire Duff's mind when he heard the roar of hunter's guns and set out to investigate who was killing the birds on his farm. But it is not too much to speculate that he thought of Seely Houk, and he probably missed him. Houk had been a principal ally in local landowners' efforts to restrict immigrant hunting. So close was Houk's connection with local farmers, and so zealous his pursuit of Italian trespassers and poachers, that even local officials had begun to wonder if he was overstepping the mark. In Mount Jackson township, several miles from Hillsville, alderman O. L. Miller complained to the Game Commission sometime in 1905 or early 1906 that Houk was exceeding his authority. As Miller explained the matter to investigators after Houk's death, the deputy game protector had informed farmers in the area that they should bring trespassers to him "and he would do the rest." Miller doubted Houk's authority "to deputize anyone in this manner," and he sent a letter to the Game Commission about it. Shortly thereafter, Miller began to notice Houk's vindictiveness toward "foreigners," who were being arrested and fined for trespassing and poaching even when they were obviously innocent.[38]

With the Non-Resident License Law to enforce and landowners to support him, Houk had considerable power over the impoverished and marginalized hunters he arrested, and for these people the very sight of him must have instilled a sense of dread. Striding through the thick woods along the banks of the Mahoning in knee-length black boots, a long, dark coat, a badge in one pocket and a pistol in the other, he represented a particularly harsh arm of the state.[39] In the bare-bones livelihood of the limestone quarries, a $10 hunting license was simply beyond the means of many people, and fines for poaching were a tremendous hardship.

Not surprisingly, Houk earned a lot of local hatred. Five or six weeks before he disappeared, alderman Haus recalled for detectives, Houk had brought an Italian poacher to trial. He was fined $25 and costs, the standard fine for violating the Non-Resident License Law. The man produced the money, and Houk left the room to find change.

As the alderman recounted, while Houk was out of the office the Italian man warned "that he would kill that s—— of a b—— before long." Such sentiment about Houk was not unusual: in fact, Haus could not immediately recall who the Italian in question might have been, "and a search of his docket would hardly bring the particular case to light, for the reason that Houk brought a great many similar cases to him and, owing to the difficulty they had in getting the right names of the Italians, the name of John Doe was used in every case."[40]

Haus remembered that two brothers, Rocco and Antonio Catalano, had threatened to kill the game warden after Houk had arrested them some years before. The lead was promising, and it seemed even stronger when one quarry owner recalled Rocco Catalano telling him that his wife in Italy needed money, but he "could not send her any, being compelled to spend" his savings paying a fine levied by Houk.[41] Detectives turned up court records showing that a "Rocco Catalino" had been arrested and fined after pointing a gun at Houk in 1903. But the leads petered out. No other evidence was found to implicate Catalano in the murder, and it appears that he was only one of several Italian men who publicly threatened Houk.

Immigrant hatred of Houk in part mirrored local resentment of game wardens elsewhere in the state, and as the Game Commission expanded its presence in the countryside, anti–game law violence increased. In 1903, deputy game protectors first ventured into the countryside to enforce the new Non-Resident License Law; in 1904, five wardens were shot at and three were hit; in 1905 no wardens were shot, but, while arresting a suspect, game warden Frank Rowe of Wilkes-Barre "was compelled to defend himself, which he did with his fists." The suspect died that day.[42] The next year was the apogee of violent confrontation over the game laws: fourteen game wardens were shot at, seven were hit, and four were killed. Among the dead were Seely Houk and Squire Duff (whose inclusion in the statistics as a "game warden" underscores the close connection between landowners and state authorities in the hunting grounds).[43]

Houk's customary aggression and his zealous enforcement of the Non-Resident License Law probably would have caught up with him even if he had lived. In fact, some local parties had begun to take action against him through the courts at the time of his disappearance. In late 1905, the game warden was arrested and tried in New Castle for assaulting Serefano Diandrea, an Italian man in his custody. The case sprung from Diandrea's allegations, made through an interpreter, that Houk had treated him "roughly" and had kept him hand-cuffed while in the back of a buggy. While in cuffs the man fell out of the buggy onto the road. Perhaps Diandrea was particularly well thought of among some local English-speakers, but whatever the reason, the district attorney was fierce in his condemnation. As Houk received half the $25 fine levied against violators of the Non-Resident License Law, the game warden was virtually guaranteed "$12.50 from every poor devil of an Italian he can arrest." The predatory nature of such law enforcement was not lost on the prosecutor, who accused Houk of making arrests solely "for the purpose of replenishing his own treasury."[44]

A jury of Houk's peers convicted him of assault, and the game warden posted bond for his release. One month later the judge denied his petition for a new trial. Sentencing was set for late March 1906. Houk did not appear on the appointed date, leaving many to speculate that he had fled the county to avoid the penalty—a maximum $1,000 fine and a year in jail.[45]

The truth was much darker. Somebody had finally beaten the game warden to the trigger. A large stone on his feet and his arms stretched out at his sides, his long coat pulled up over his head and filled with rocks, Seely Houk had been face down in the Mahoning River for three weeks.[46]

The Society

As investigators came to see it, Houk's demise was the culmination of a sequence of events that began in 1905. In August or September of that year, a young immigrant named Luigi Rittorto went hunting for groundhogs near Hillsville. Borrowing a double-barreled shotgun and a dog from his employer, Rocco Racco, a Hillsville store owner, he headed into the woods. Rittorto did not have a hunting license. He knew from warnings his neighbors had given him that he should watch out for "the police." The same neighbors told him that if caught hunting, he would spend one year in the penitentiary and pay a $50 fine, a huge sum for a man who earned only $10 and board each month.

In the woods he saw an armed man, whom he later identified from photographs as Seely Houk, approaching from the direction of one of the quarries. Fearing that this was a policeman, Rittorto immediately began to run. He heard a shot but kept running. He must have been terrified, for in his haste his stomach brushed a barbed wire fence, tearing the flesh, and he ran through thick briars that shredded the skin on his hands. He made a detour of two or three miles to avoid going through the woods where he might be apprehended. By the time he arrived home, he was bleeding profusely. Rittorto told his employer, Rocco Racco, what had happened. The older man assured him that he was safe now that he had reached home. Nevertheless, the young man was so frightened that afternoon that he could not eat, wondering if at any moment the game warden might walk through the door and arrest him.[47]

Detectives took a special interest in what purportedly happened next. The evening the hunter returned, his employer noticed that one of his dogs was missing. Rittorto confirmed that he had taken the dog hunting and left it when he fled from the game warden, who he now suggested may have shot the animal. Either that night or the next day, Racco found his dog in the woods, dead from a gunshot wound. Later, witnesses testified that Racco threatened to kill Houk in retribution for the dog's death.[48]

The incident supplied prosecutors with a motive for Houk's murder. To the non-Italian public, the story reinforced assumptions about the dangerous passions of dark-skinned immigrants who would kill a man over a dog. But ethnic and racial stereotypes aside, many people, Italians and non-Italians alike, had no difficulty envisioning Racco making such a threat or acting on it, for Rocco Racco was more than a store owner. He was also the founder of several organizations in Hillsville, among them a shadowy clique known to its members as the Society or the Association. To authorities and the larger public it had another name: *mano nera,* or the Black Hand.[49]

At the turn of the century, the drawing of a black hand print on threatening messages—extortion demands, ransom notes, and so forth—was a pervasive practice among criminals, mostly but not entirely in Italian communities. The Hillsville organization was just one of the myriad organizations that officials associated with this practice; American commentators used the term "Black Hand" as a euphemism for virtually all Italian immigrant crime until the 1920s, when it was succeeded in official parlance by "Mafia." Supposedly all Black Hand organizations were part of a giant criminal syndicate, a monolithic transatlantic organization, under the thrall of crime bosses in Italy and bent on undermining American government and American society.[50]

To understand how Racco and his associates might have been connected to the Houk murder, their organization must be viewed within the social context of the Hillsville settlement, uncovering its meanings for local immigrants. The term Black Hand is of little use for this purpose. Contemporary commentators applied it to so many distinctive activities and associations, criminal and legitimate alike, that it became a way of stigmatizing Italians for almost any activity. Although non-member immigrants certainly called them Black Hand, rival factions within the organization were unanimous in attesting that they never called themselves by that name.[51] There is no reason to doubt them on this point: despite its public currency, the name Black Hand was rarely used by Italian gangs. It was not of Italian origin, and even if it had been, it was so easily pirated by any ambitious extortionist as to detract from the kinds of monopolies on intimidation that criminal syndicates sought.[52]

What was the Hillsville "Society"? As secretive as it was, the extraordinary documentary record of this case nonetheless affords some intriguing insights. Most interesting are the daily reports of one undercover agent of the Pinkerton Detective Agency. Known in the records only as Operative #89S, or by his initials, D.P., he was dispatched by his superiors to Hillsville in the summer of

1906. There D.P. rented a bed at a boarding house and became a quarry laborer. He remained in Hillsville for about a year and a half. During that time he became a leading member of Racco's organization. Every day he wrote summaries of his activities, and every few days he mailed them to his Pinkerton superiors.

This correspondence is part of a tortuous trail of sometimes murky evidence, an investigation fraught with pitfalls, in which the workings of Racco's mysterious "Association" appear through the accounts of an anonymous detective. The source's employment by the Pinkertons urges caution in interpreting his accounts. Pinkerton agents were not objective observers. They were often mercenaries for union-busting corporations, and among workers their ruthlessness earned them a reputation as thugs and even killers. As detectives, they could be notorious liars.

Yet D.P.'s observations command attention. They were inevitably detailed and mundane, seldom subject to the grandiose claims of other Pinkertons working the case. Most detectives joined officials and the popular press, seeing the nefarious Black Hand around virtually every corner in northwestern Pennsylvania. By conjuring up the specter of an international criminal conspiracy, publishers sold newspapers, and state authorities and Pinkerton supervisors drummed up public support for their Black Hand investigations.

In this context, a striking feature of D.P.'s reports is his near-total avoidance of the Black Hand name in favor of more banal monikers, the Society or the Association. D.P. alone reported that the society leadership frowned on use of the black hand symbol, preferring to make their threats in person.[53] With their frequent references to "the Black Hand," authorities during the Houk murder investigation suggested their unfamiliarity with the inner workings of the Hillsville gang; in avoiding it, D.P.'s accounts suggest a high degree of reliability.

Other factors reinforce his credibility. His careful rendering of daily routine—"This morning I went to work at 7.00 A.M. I loaded cars and drilled stones until 12.00 noon when I left and went home for dinner"—hardly suggests a fevered imagination at work. Perhaps the greatest testament to D.P.'s veracity was his official failure: despite what must have been tremendous pressure from his superiors, D.P. never connected Racco to the Houk murder. In fact, he defied his superiors and suggested—strongly—that the state's investigation was leaving out a crucial suspect. Read carefully, his daily reports go beyond official blindness and establish what roles this mysterious Society played in the local community.

On this point, D.P. was unequivocal: it was "generally understood" by immigrants "that the Society is what is known in Italy as the Mafia."[54] On its face, this could be read as a sensationalist allegation, as the term *mafia* has been as widely and incorrectly used as Black Hand ever was. But mafia, unlike Black Hand, is a genuine historical phenomenon. Although there is still disagreement about what defines an organization as a mafia, most scholars concur that the term best describes a complex of socioeconomic and political behavior, a way of structuring an industry, that originated in southern Italy for particular reasons. The principal activities of mafia firms generally involve the selling of "protection" in many ways. In its most exploitative form, mafia protection is a euphemism for extortion, but, historically, successful firms or families (the firms usually have a core of blood relatives) operate a broad range of more sophisticated businesses. These include dispute settlement, for example, in which opponents agree to abide by a contract, witnessed and enforced by a presiding mafioso. Labor contracts can be enforced in similar ways. Often mafiosi serve as guarantors of products in a market exchange: when purchasing a horse or a wagon from an unknown salesman, a buyer might call on a mafioso to guarantee the sale. If the product is shoddy, the mafioso's witnessing of the sale guarantees that the buyer can return it. At the same time, sellers might acquiesce to mafia involvement in their business as a way of ensuring payment for goods: if the buyer defaults, the mafioso will collect.[55]

It is only in the past decade that scholars have come to understand the phenomenon of mafia as the business of selling protection. In 1906, few if any American commentators could have characterized it thus. D.P.'s letters are therefore all the more remarkable for describing an organization, Racco's Society, dedicated to the protection industry, in its organization and modus operandi bearing close resemblance to southern Italian mafia in general and the mafia of Calabria (the home province of most Hillsville immigrants) in particular.

Racco's family members were the core of the organization in its early days, but the Society required a large number of members for a wide variety of occupations, and toward this end recruitment of new adherents was a constant concern.[56] Most mafia organizations require the swearing of a blood oath and a highly ritualized, ceremonial induction for new members. Mafia firms in Calabria are peculiar in requiring an oral exam. Petitioners for membership in the Hillsville Society also had to pass an oral exam, apparently akin to Calabrian mafia tests in which initiates recount bandit folklore and evoke legendary figures.[57] Judging from D.P.'s accounts of his own examination and induction, many immigrant men were joining the organization in 1906. Undoubtedly they were compelled by its monopolistic power across the entire range of Hillsville economy and society.

In the workplace, the Society served as a power broker between quarry owners and workers. At least one Society member served as a labor contractor, bringing on extra laborers when needed in return for a fee from the owners, and other leaders gave special concessions to members. In return for a cash payment, Racco would move a favored underling into one of the better-paying quarry stations; he also exploited a position of minor authority in the quarries to obtain unearned pay for cohorts.[58]

In return for their loyalty and dues payments, the Society also helped members in the event of unforeseen debts, illness, and injuries. Services such as guaranteeing board payments for new arrivals and credit payments in time of misfortune were probably indispensable to many, and the Society paid compensation to sick, injured, or disabled members.[59] In all likelihood, these benefits were not as extensive as they might initially seem, and, like mafia in southern Italy, the Society in Hillsville often undertook a benevolent cause as a ruse to secure members' money for the leaders' own use.[60] Yet even the slightest compensation for illness or injury, the least help with obtaining and keeping a good job, was more than most immigrants could expect from their employers or the government. As one Pinkerton operative reported from Hillsville, to some members the Society represented more than a criminal organization. "The black hand society state that the society is for the good of its members, like a labor union and its inferior members believe this."[61]

Racco's purported associations went beyond the workplace and the town, even beyond this world. Like mafia headmen in southern Italy, Racco took a leading role in the town's religious life. Even after he lost his place at the head of the Society, he remained president of both major church groups in Hillsville, the St. Lucy Society and the St. Rocco Society. These positions required that he act as chief sponsor of the town's annual church festivals, a role that no doubt required him to lay out significant sums of money while giving him the opportunity to showcase his magnanimity and status, all the time implying his ability to provide holy protection for clients.[62] And, like his counterparts in the homeland, Racco was godfather to dozens of children in the area, making him a central figure in local families and a kind of mediator between them and God.[63]

The Society maintained its operations through classic methods of domination, monopolizing links between Hillsville residents and the outside world.[64] It dominated not only the quarries, where most immigrants made their money, but to some degree it dictated where workers spent their earnings. Society leaders maintained a virtual monopoly on beer sales in Hillsville, and while Rocco Racco was president, all members were required to buy their groceries from his store.[65]

Perhaps the Society's most valuable tool was the monopoly it attempted to keep on criminal and criminally violent behavior. Illegal activity had to be sanctioned by the Society or the perpetrators could face severe penalties, as when D.P. reported that "lots of people paid $200 and $300 for stealing chickens."[66] Society leaders preferred that violence also remain under their control. After mediating an altercation in which one man was stabbed, one of the Society's leading figures warned even D.P. against using personal violence to settle a score. "He recommends me to behave myself, and not do any fighting, but to leave everything to the association to settle, and they will punish the one that needs it."[67]

The Society asserted its pervasive influence in these diverse social arenas with the orthodox technique of control in mafia organizations: terror. The Society, and in particular Rocco Racco, embodied extreme violence. Even after he was no longer a ranking Society member, Racco enthusiastically led armed forays into nearby Lowellville to kill immigrant men who had publicly slighted Society adherents.[68]

Locals felt the Society's sting often, for the organization's most prominent activity was extortion of Hillsville's quarry workers. Any Hillsville man was a potential Society "member," insofar as he might be made to pay "dues."[69] Those who refused to make payments might face the ordeal of Nick Ciurleo. Having repeatedly refused to pay a $100 "initiation fee" into the Society, he endured a beating by dozens of Society men who then took turns spitting on him before they painted him head-to-toe with a broom dipped in a bucket of excrement. Other men acquiesced to the payment of weekly dues in varying amounts but never attended meetings or became involved in Society activities; they were thus "members" only insofar as they were buying protection from the type of treatment meted out to Ciurleo.[70] When one Society leader told an investigator that almost all the Italians in the Hillsville vicinity were Society "members," he was probably not exaggerating.[71] When protection from the Society could be acquired only by joining, men joined.

Given Houk's violence and the predatory nature of his law enforcement, it may well be that some in the community sought Society protection from him. But evidence suggests a more complicated story. As extensive as Society influence was, it is ironic that the collision with Houk resulted partly from the Society's failure to control one source of immigrant livelihood: the hunting grounds. In no sense did the Society seek to manage local wildlife or regulate access to it. Local Italians hunted where they chose, when they had the time, and when they thought they could avoid prosecution. As there was no money to be made from subsistence hunting, and as it initially attracted no attention from legal authorities, Society leaders saw no reason to intercede in it—at first. Their interests were drawn more to extortion and bootlegging, activities which were both lucrative and highly illegal.

Only when conservationists attempted to keep Italian hunters from wildlife did the Society become concerned. For the leadership, control over the immigrant community was always tenuous and incomplete: immigrants moved in and out of town constantly. Establishing complete dominance over such a transient settlement was vastly different from controlling sedentary peasants in southern Italy. And there were other threats, too, not least the numerous competing "associations" and "societies" in surrounding towns who constantly sought to undermine the Hillsville Society. With Houk arresting and fining dozens of local men, the Society's monopoly on intimidation became more porous than ever. Effectively, Houk's unstinting enforcement of the game laws erected a barrier between local immigrants and part of the natural world upon which they relied for subsistence. For an organization that survived because it dominated critical links to food, money, and employment, this was a significant challenge.

And it is not too much to speculate that some Society men saw a still dimmer future with Houk in the community's environs. The Society might consolidate support if the mere sight of the "game policeman" could send Italian hunters like Luigi Rittorto running to the protection of Society leaders like Racco. But what would happen if someone, someday, ran to the "game policeman" for protection from the Society? Houk's shooting of Rocco Racco's dog could only have underlined for Society leaders that they could not long evade a confrontation with him. In leveling his gun at the game warden, Houk's murderer—whoever it was—probably aimed to shore up Society dominance at least as much as to protect local hunters. A supervisor in the quarry recalled that the day Houk's body was found, three Italian workers "were very curious to learn who it was and on learning that it was Houk, they lost all curios[i]ty."

Although the supervisor did not believe that these men had anything to do with the murder, it seemed "they were expecting something of the sort to happen."[72]

It might seem that immigrants like these gave their loyalty to the Society out of some sense of shared community. In truth, Society leaders like Racco were hated as much as Houk. After Racco's arrest, D.P. reported, "Of all the people who know Rocco Racco, there is not one to say a good word for him, and they do not care for him at all." Hillsville people generally seemed "very little interested in the future welfare of Rocco Racco."[73] Society domination of Hillsville, maintained as it was by terror, did not breed affection for Society leaders. It survived because Hillsville Italians were new to the United States, unfamiliar with its laws and their own legal prerogatives, largely unable to speak English, and often isolated from friends or family, and because the community was transient and Society exactions were probably seen as temporary difficulties. For Society leaders, the presence of an aggressive police authority was intolerable precisely because many immigrants turned to the state the minute they thought it willing and able to crush the Society. Houk's killing of Rocco Racco's dog became widely known in the area, and not just among Italians. Alderman Haus warned the game warden to be extremely careful lest "the Italians hurt him." A local Italian claimed that Houk had told him, "Yes, I killed Rocco Racco's dog, but I am not afraid of him killing me; if he is any quicker on the trigger than I am let him try it."[74]

Whether or not anyone threatened Houk, nothing happened to him for some time after he killed the dog. But in the early months of 1906, events within the Society made it an increasingly unstable organization prone to sudden violence. Sometime before spring 1906 the unity of the Society's upper echelon—Racco, Ferdinando Surace, and Giuseppe Bagnato—dissolved into a fractious contest for the organization's leadership, with Racco on one side and Bagnato and Surace on the other.

The origins of this split are murky, but they involved a strike in February, during which the quarry owners—acting in the best tradition of landowners in Italy—hired the local mafia to protect company property from violent strikers. Two dozen members joined Racco in defending the company, and Racco himself acted as a watchman.[75]

Mafia families are notoriously anti-labor, except in Calabria. In the home province of most Hillsville immigrants, there is a tradition of mafia-worker alliance in non-union labor action, perhaps because the absence of a union allows mafiosi an opening in which to broker power between employers and workers.[76] This may help explain what happened next in Hillsville. Surace and Bagnato sided with the strikers. They excoriated Racco for agreeing to act as a company watchman, labeling his action as treason and tantamount to coopera- tion with the police. In an event that may have been related, Surace leveled allegations of sleeping with another Society member's wife against Racco. Such charges required that Racco face a Society trial and be executed if guilty.[77] True or not, Surace's accusations against the Society president suggest that he had a powerful following of his own within the organization. By the end of February 1906, Racco's position was in serious jeopardy.

In an organization where men proved their ability to lead in large part through their willingness to dispense violence, a contest for leadership was potentially explosive. Although the fight between Racco and the others did not move into the open until mid-March, their competing ambitions most likely helped to put Hillsville through its bloodiest passage in the months leading up to that time. The *New Castle News* reported that from the end of 1905 into the beginning of 1906, "men of all classes were attacked by Italians near Hillsville," and many of them sent to nearby hospitals "with throats cut, faces slashed and bearing stiletto wounds." Attempts by local police officers to resolve the vio- lence met more brutality: county detective McFate was wounded in a shootout when he arrived in Hillsville to investigate Houk's disappearance. Outsiders obviously agreed with the Society headman who remarked in September 1906, "there has been too much cutting and killing up here for the past year."[78]

Within the Society, Racco's grip was slipping rapidly. Mafia leaders from neighboring towns and from as far away as New York arrived in Hillsville at Surace's request, and a trial was held on March 10, 1906. Racco was found guilty of adultery and sentenced to die, but he apparently managed to have the penalty set aside after he paid Bagnato $500 and lesser amounts to Surace.[79]

Seen in the context of a struggle for control of the Society, Houk's death begins to seem almost predictable: as threatening as he was to mafia interests, with no local connections to either of two competing factions, he probably became a potential target of both. Sources from within the Society eventually told investigators that in March 1906, Racco fulfilled his promise to kill the game protector. Allegedly, Racco hid in the woods with his brother-in-law, Vin- cenzo Murdocca, waiting for Houk. Both men carried double-barrelled shot- guns. Murdocca fired a shot to catch the game warden's attention, whereupon

Houk entered the woods, and Murdocca shot him in the face. Immediately thereafter, Racco stepped up to Houk and fired two shots into the game warden's body. Later the same day they put the body in the river.[80]

Racco's desire to retain power could have motivated him to kill the hated warden. He had been accused of cooperating with the police during the strike; what better way of countering such charges than to kill "the game policeman"? His brother-in-law fled the country not long after Houk's disappearance.[81] Certainly the parallels between this story and the known physical facts of the murder lent the accusations credibility, and they proved damning in court.

Yet they also ring like something less than the full truth. The description of the murder came from none other than Ferdinando Surace, the rival who by the time of the murder trial had replaced Racco as president of the Society. Surace testified that Racco invited him to his cellar one cold spring day in 1906. After drinking some wine together, they went back outside, where Racco said, "You must swear to me you will never say anything if I tell you something."[82] Surace duly swore, and Racco told him the story of how he and Murdocca had murdered Houk a day or two before.

Straightforward as it was, this testimony leaves a number of questions unanswered. Seely Houk was killed on March 2, 1906. By that time, Racco and Surace were feuding for control of the Society. Racco's Society trial for adultery—on charges that Surace had lodged against him—was only days away. Why did Racco invite him in for wine? Why did Surace accept such an offer? Racco might have made a boast of extreme violence in order to intimidate Surace, but the context of the conversation—the men had finished their wine and were standing by a water pump in the sunshine, talking like two neighbors—makes Surace's story dubious at best. Add to this the fact that investigators only extracted it from Surace after he had been in jail for months awaiting his own trial on various charges, and it seems even less likely that it contained the whole truth. Surace's underlings, in jail with him, confirmed the basic outline of his testimony, but Surace's cellmate botched his testimony and reported that Racco confessed to killing Squire Duff, not Houk.[83] It seems most likely that Surace and his followers had concocted a story about who killed Houk in order to appease aggressive officials, to garner lighter sentences for themselves, and not least to dispose of Racco.

Other sources also fail to corroborate Racco's guilt. Surace's credibility at the trial was in part attributable to rumors, supposedly widespread in Hillsville, that Rocco Racco had killed the hated game warden.[84] But other rumors told different stories. At least one highly placed Society member, Salvatore Candido, told authorities that Surace must have had something to do with the murder, and another Pinkerton operative reported rumors that Bagnato had killed the game warden.[85] The rumor mill was simply reflecting the popular experience of Hillsville residents: Houk's murderer would be found among men who were capable of such abhorrent behavior, namely, the Society leaders. For his part, D.P. downplayed rumors about Racco being the murderer and suggested that the brother-in-law, Vincenzo Murdocca, was a much more likely suspect.[86]

Racco's murder trial was a sensation, well-covered in the press and mobbed by city residents. The director of the Pennsylvania Game Commission, Joseph Kalbfus, even came to town to watch the trial. As the jury deliberated, he delivered a lecture in the New Castle YMCA. In a sense, trial and lecture both expressed deep-seated fears of a rapidly changing countryside. In deciding Racco's guilt or innocence, the jury did more than resolve a criminal trial. They also acted to reassert civic power over a community that seemed alien, violent, and disruptive of rural society and culture.

If the trial evidence inspired anxiety about the future, Kalbfus's message was even more fearful. His lecture at the YMCA focused on the ecological importance of birds. Without them, insects would devour all vegetation in America within ten years. And the poachers of Hillsville were no small threat to birds. Indeed, immigrant poaching represented nothing less than "the greatest menace to our bird life."[87] Kalbfus's condemnation of immigrants in general found an echo in the jury's verdict of the immigrant before them. In October 1909, Rocco Racco went to the gallows for the murder of Seely Houk.[88]

Local People, Local Authority, and State Law

On March 2, 1906, Seely Houk rode the streetcar out of New Castle, heading for Hillsville. Asked if he was returning that evening, the streetcar driver remembered the game warden telling him, "His life would not be worth a cent, if they saw him at night time up there."[89] Disembarking at Kennedy's Crossing, Houk met a prior acquaintance, a local farmer's daughter named Nora Martin, who had just put her sister on the car to New Castle. Martin prevailed upon the game warden to ride the rest of the way to Hillsville in her wagon, as she was going home that way.

On the way, Houk mentioned "that he was looking for two fellows, one of whom he named." But the woman, like so many others, had no ear for Italian names and quickly forgot these. She asked to see the warden's handcuffs; Houk obliged. It was a little after five P.M. when the two reached the farm of her father, Silas Martin. The game warden helped the farmer put the horses away, then stayed to talk for ten minutes. Houk mentioned that he was "going to keep a sharp watch" on the Hillsville quarry workers, "as the evenings were long now and the men would be hunting after supper." He had warrants for the arrest of two men and wanted to catch them as they left work that day. The game warden walked off through the fields, toward the Johnson quarry. It was almost sunset.

From physical evidence, investigators pieced together what happened next. As evening came on, Houk entered the woods, probably on his customary patrol route. Thirty feet from the warden, a shotgun blast ripped the air, catching him in the face, neck, and chest, and catapulting him onto his back. An assailant stepped up to the prostrate form, lowered a shotgun loaded with No. 6 birdshot to Houk's mouth, and pulled the trigger.[90]

This horrible vignette became a legend at the Game Commission, where Seely Houk became a conservationist martyr to the Italian poaching menace. The hanging of Rocco Racco signaled the close of the Game Commission's troubles in the New Castle area. With convictions of dozens of Hillsville men for mafia activity, the Society was broken. Bagnato had fled, Racco was dead, and Surace was in jail.[91]

Like all mafia organizations, the Society functioned best when it was largely unified and where state authority was weak. Internal rivalries eroded leaders' ability to "protect" local residents, who found themselves subject to extortion demands from competing leaders.[92] The organization could not operate in a setting where undercover detectives became trusted cohorts. Its monopoly on intimidation slipped as the state mustered greater and greater force, with a contingent of mounted state police temporarily occupying Hillsville during the "Black Hand trials" of 1907, which sent many Society men to prison.[93] The shooting of the "game policeman," rather than securing Society control, brought its downfall.

In all likelihood, only some of the armed confrontations between Italian immigrants and game wardens involved gangs like the Hillsville Society. The onerous burden of the game laws probably motivated some poachers to take a chance on shooting at "the game policeman" rather than face the terror of jail in a strange country. The three Black Hand death threats that arrived at the Game Commission in the early 1900s were probably the work of disgruntled individuals, because few if any genuine mafia organizations used the symbol. But the arrival of state authority in the hunting grounds, of badge-carrying, armed policemen to monitor the widespread hunting which characterized Italian immigrant settlements, represented an inadvertent challenge to mafia authority. This fact does much to explain the pattern of violent response to the game laws in Italian immigrant communities, where mafia political structures arrived in the cultural baggage of immigrants, transported alongside many more humane and benevolent traditions. Like those traditions, it was reworked and applied pragmatically to its new American home. As it matured, American mafia took on a very different appearance, in some respects, and the reasons for the changes remain unclear. The connections between the urban Sicilian-American mafia families of the Prohibition Era and the rural Calabrese immigrant associations of the early 1900s await further research.

After 1907, the Game Commission reported no more Italian attacks on game wardens.[94] In 1909, largely as a result of the Houk case, the Game Commission secured passage of the Alien Gun Law, banning hunting and the ownership of firearms by non-naturalized immigrants. The state timed a special show of its enforcement: the day Racco was hanged, authorities arrested eight Italian hunters in the Hillsville environs. Five paid their fines, the others sat out their time in jail. All forfeited their guns.[95] Because it was illegal to hunt with any weapon but a gun, this new measure effectively pushed non-citizens out of the state wildlife commons almost completely, leaving them the recourse of fishing and trapping. Indeed, the commission reported (somewhat optimistically) that by 1910 it was "a rare thing indeed . . . to receive a complaint charging an alien" with a game law violation.[96]

The story of the Italian hunters in Pennsylvania suggests how local struggles over hunting resulted in the imposition of state power over certain local communities. Immigrant efforts to shape the local commons, to make local wildlife accessible to themselves within a familiar cultural context, were tied closely to immigrant community organization and structures of power. Immigrant ideas of the commons clashed with the older, resident common property system, challenging native locals' notions of landowner prerogative, and exacerbating social tensions. Local citizens, in defense of their own notions of what was proper behavior in the hunting grounds and who should be permitted to hunt at all, called on the state for assistance. In this way, the state gathered power and furthered the process of converting local game to public goods under state protection.

The story of Racco, Houk, and the Society came to dominate the history of immigrant experience with Pennsylvania's game laws. But its more subtle implications are discernible in the story of Racco's employee, Luigi Rittorto, whose hunting trip in 1905 ostensibly began the dreadful sequence of events. On that day when he fled Seely Houk, Rittorto typified the travail of numerous immigrants who relied on subsistence hunting for food. Such people were frequently without recourse to legal counsel and without means to pay the heavy fines imposed on them for breaking the game laws.

But, while resistance to game laws was a virtual necessity for immigrants like Rittorto, the young man's cooperation with detectives was also typical. Rittorto was eager to testify against his former employer, and he told investigators "that he was glad that Rocco Racco and the rest of his band who were members of the Black Hand Society were in the Penitentiary; that since he has been there Hillsville has become a Paradise."[97] Immigrants, however marginalized, could see themselves as a segment of local society. Often, they were able and willing to separate mafia from their own lives and to help state authorities reorder social structures while they continued to resist state authority in wildlife. In his mixing of resistance to and alliance with the state, Luigi Rittorto illustrated the complex layers of local response to the emergence of the state commons in wildlife.

Only the testimony of immigrants like Rittorto allowed state authorities to defeat Hillsville's Society. In one sense, immigrant strategy was similar to that of non-Italian locals, who demanded state action to end the power of gang leaders in the small town. But unlike other residents of the Hillsville area, many immigrants never accepted state game laws that did not allow non-citizens to hunt. Resistance to the game laws became less violent and more covert. For decades after Racco's execution, the Game Commission reported nearly 200 arrests per year for violations of the Alien Gun Law.[98] If these figures are any indication, immigrants gained the commons they sought by going underground, their activities more than ever characterized by stealth, not violence.

Hillsville residents recall that, until the 1930s, non-naturalized immigrants frequently hunted for groundhogs, birds, and rabbits in the fields and forests surrounding the town. Farmers did not object so long as no property was damaged and the poachers remained hidden. They learned to hunt "on the sneak."[99] To some degree, accommodation was in the landowners' interest. If the Italians did not work the quarries and the mills, who would? A sort of tacit agreement between farmers and hunters emerged, allowing local men access to the hunting grounds, in violation of state law but in keeping with local understandings of landowner prerogative and laborer use rights. This unspoken concordance in essence defined the boundaries of a new local commons, embedded within the state commons and operating parallel to it.

With the maturation of American-born children who were citizens, the growth of successful citizenship applications, and the improvement of wages during the 1940s, hunting of groundhogs, songbirds, and other subsistence animals began to disappear in immigrant communities. Children of immigrants joined other American hunters and turned increasingly to recreational deer hunting, which became as much a rite of passage as sandlot baseball for this generation, "the first Americans."[100] Many helped form a new organization in Hillsville and nearby towns: the Mahoning Sportsmen's Association, which began in the 1930s as an effort to support game propagation efforts, gun safety, and "sportsmanlike" hunting. Today, this association has a well-appointed clubhouse with a rifle range, a swimming pool, and a membership of 2,000. The bass teeming in many of the flooded, abandoned quarries that dot the countryside testify to their efforts at stocking the land for sport. Among subsequent generations of Italian descent, the early anti-Italian game laws became curious relics.[101]

But to their predecessors, the non-citizen immigrants who poached quietly in the farm fields and hills around Hillsville, there was no doubt that the laws imposed a great hardship. Ensuring welfare of self and family meant an occasional trip to the fields for a groundhog, a rabbit, or a brace of sparrows. That being reality, the state authorities and their game laws deserved little attention beyond watchfulness, a skill cultivated among hunters generally, but especially so in Hillsville.

2

Boon and Bust

Pennsylvania's Deer Among
Sportsmen and Farmers

In 1889, Pittsburgh industrialist John M. Phillips went hunting with a friend, Hiram Frost, in Elk County, Pennsylvania. Shortly after daylight, between the towns of Ridgway and Brockwayville in six inches of snow, they flushed a deer. They trailed the animal all day, camped on the trail that night, and tracked it all the next day. At dawn on the third day they took up the pursuit again. Soon thereafter, they found the buck, and Phillips shot it. As Phillips later recalled, "During all that long chase we did not cross another deer track. As we stood over the dead buck I said to my friend: 'I fear I have killed the last deer in Pennsylvania. I will never kill another in this state.'"[1]

Phillips's fears about deer extinction were widely shared; by the turn of the century, many recreational hunters complained that deer numbers were rapidly declining. Phillips, along with many of his hunting colleagues, later sought a political solution to the problem of game scarcity. Hoping to change the landscapes of Elk County and the rest of Pennsylvania into a countryside of abundant deer, he went on to become a founding member of the Pennsylvania Sportsmen's Association and a strong advocate for the Pennsylvania Game Commission.[2]

Less than forty years after the Phillips hunting trip, biologist Vernon Bailey confronted a very different Elk County. Commissioned by the state of Pennsylvania to investigate deer conditions, Bailey found "excessive numbers" of deer in Elk County and three neighboring districts. Hungry deer were stripping the range, resulting in "the partial or almost complete destruction of certain species of plants, shrubs, and trees." Authorities planted seedlings to restore deer habitat, but the deer ate almost all of them before they could mature. In Elk County and elsewhere, the animals were invading the fields of outraged farmers. They were "very fond of many crops including corn, buckwheat, oats, beans, cabbage, potatoes, sweet potatoes, apples, and grapes. In fact," Bailey went on, "there seem to be few farm crops which they do not relish."[3]

The changes in Elk County's ecosystems were typical of many others across Pennsylvania in the early twentieth century. Where there had been few deer in 1900, by the 1920s there were many. This biotic shift occasioned a political crisis. Conservation had been a political and social response to a landscape of scarcity. By placing the hunting grounds under state control and making deer a common pool resource managed by the state, authorities attempted to create a landscape of game abundance. When they succeeded, few celebrated. Deer ravaged agricultural crops, stripped forests bare as high as they could reach on their hind legs, and still starved in such numbers as to make rural valleys putrid with the stench of rotting carcasses.

The situation drove a wedge between erstwhile conservationist allies. Since the end of the nineteenth century, rural landowners and urban sportsmen had combined to create a conservationist consensus. When landowners sought control over out-of-state hunters to protect farm property in 1901, the result had been Pennsylvania's first hunting license law. Through the efforts of the Game Commission, urban sportsmen like Phillips lent their support to farmers in Hillsville and elsewhere. When the state was unable to fund an extended investigation of the murder of deputy game protector Seely Houk, Phillips solicited private donations to hire the Pinkertons.[4] Sportsmen were more interested in protecting animals than farms, but because regulating hunters could achieve both ends, farmers and recreational hunters joined causes. Conservation at the turn of the century was an alliance of rural and urban elites arrayed against more marginalized rural people, among them immigrants and—as became clear in the case of deer hunting—residents of the northern mountain counties.

The deer irruption that Vernon Bailey described in 1927 drew new lines between city and country, pitting landowners against urban recreational hunters in a struggle to redefine "proper" uses of deer. In this dispute, farmers often assumed the mantle of beleaguered locals, battling an insensitive, remote Game Commission. Insofar as deer were public property in which most Pennsylvanians (but not unnaturalized immigrants) had a limited bundle of use rights, the animals were a "state commons," a common pool resource managed by the state on behalf of the Pennsylvania public. What chagrined local farmers was that this state commons was largely divorced from place—the state owned practically all deer, on private and public land, and managed them essentially as one giant resource. The state exercised its power to stop farmers from killing deer even in their own fields; in calling for more power to determine their own interactions with game in local areas, farmers were demanding a limited return to the local commons.

This dispute had several other dimensions, among them competing ideas of culture and economy. The basis of rural livelihood—the need to produce crops for market—often clashed with urban sportsmen's aesthetic attachments to deer. To most farmers, a female deer was a ravenous, crop-wrecking animal who gave birth to more of the same, but to most sportsmen the same animal was a delicate symbol of femininity and motherhood.

Contemporaneous with the struggle to define cultural meanings of wild animals was a battle to define the place of markets in the hunting grounds. Lower-class and poorer farmers were frequently among those who sold game meat for cash. Blaming market hunters for the decline of many species, conservationists outlawed market hunting and banned commerce in game. Some rural people defied state laws that prohibited selling game meat, but wealthier farmers sided with urban hunters on this issue in order to better control the depredations of hunters on their lands.

And yet, as much as the creation of a state commons in game entailed a contest between anti-market and market forces, state power and market function were not mutually exclusive. Indeed, conservationists relied on market forces to assert state power over the hunting grounds. Combatting the market in game required the construction of a state-dominated, heavily regulated market in the activity of hunting itself. The market that emerged to replace the older game market was a peculiar one, with many limitations, but its chief effect was to give the state enormous economic and political power in regulating access to wildlife.[5]

Nevertheless, all the state's political power could not forestall the changes in Pennsylvania's ecosystems that occasioned this crisis. In many ways, rural and urban interests alike were seeking answers to one simple question: how best to respond to protean countryside, bereft of deer one year, teeming with the animals another?

To understand the magnitude of this change, one must begin with the paucity of deer in Pennsylvania at the turn of the century. By the second half of the nineteenth century, there were few deer anywhere in the American northeast. Phillips's hunting story was part of a genre of oral traditions from the late nineteenth century, stories of abundant hunting grounds laid waste. Joseph Kalbfus, head of the Pennsylvania Game Commission and a close friend of John Phillips, grew up in Maryland and recalled in his memoirs that he had never seen a deer until his late teens, when he went to Colorado.[6] J. Q. Creveling, an avid outdoorsman and pigeon hunter in Pennsylvania in the late 1800s, wrote that as of 1900 he had seen a deer outside of a circus only once, and that one was a tame animal on a farm.[7]

The late nineteenth century was a period of low deer populations for two reasons. First, hunting for subsistence and market was widespread. Hunting was integral to farm life, and many people in Pennsylvania lived on farms. Pervasive, year-round hunting restricted deer populations to areas that could provide food, dense cover, and distance from human populations. Thus, Henry Shoemaker's 1915 survey of famous Pennsylvania deer hunters was dominated by residents of the north-central counties, where sparse human settlement guaranteed relatively light hunting pressure, and rugged mountains and thick hemlock forest provided adequate cover and forage for deer.[8] As is usual among subsistence hunters, lifetime bag totals for northern Pennsylvania hunters were high by conservationist standards. For example, Edward H. Dickenson of Potter County, who lived to be eighty, killed 1,100 deer in his lifetime.[9] Beyond the north-central counties, deer were less able to survive. Any that might wander into the flatter farmlands—where human settlement was denser—were more likely to fall prey to hunters' guns. If farmers beyond the mountain country killed fewer deer than Edward Dickenson did, it was not because they abstained from killing as many deer as they could. It was because there were fewer deer in the vicinity.

The second and perhaps more important reason for the paucity of deer lay in the changing relation between Pennsylvanians and their forests. In the northern counties and elsewhere, the farm economy had always depended to a degree on forest products. Creating fields to plow required clearing them of trees, which often became fences, furniture, houses, and fuel. Some wood could be sold as charcoal or potash. The rest was either burned or processed in local saw mills for sale in cities.[10]

When hunting pressure was not too heavy, nineteenth-century settlement left room for deer habitat. White-tailed deer favor "edge" environments, where they can forage on the dense ground cover of clearings, but escape from danger or inclement weather into nearby forest.[11] Farm fields amid the forests provided ideal deer habitat, albeit with considerable threat from hunters.

The viability of deer habitat declined as people developed new uses for the forests. By 1850, much of Pennsylvania's remaining old-growth woodlands consisted of white pine and hemlock stands. White pine, prized for its straight boles, was soon cut even from relatively remote mountainsides. The hemlocks remained, largely because the wood was too brittle to be used as commercial timber and the hemlock forest lands, in the north-central counties, were ill-suited for farming. But the invention of the wire nail in the 1860s made hemlock usable as a building material, and the expansion of the tanning industry after the Civil War made tannic acid—an extract of hemlock bark—a valuable commodity.

Combined with the increasing centralization of milling, these factors brought trees of the north-central counties into a new relation with society, one in which the consumers of forest products were more extra-local than local. Once, local entrepreneurs without heavy capital had shaped the composition of the forests, clearing land for farms or selling timber in local markets. Now lumbermen from Williamsport joined with their counterparts from Maine to dominate an extractive enterprise that brought hemlock trees from the north-central counties down the Susquehanna River to Williamsport. There large mills and woodworking firms converted them to timber and finished products, including tannic acid and wood chemicals for eastern markets. By 1880, the remaining hemlock and mixed hardwood forests were being funneled into a corporate processing and marketing system that made Pennsylvania the nation's primary logging state.[12]

That position deteriorated rapidly: by 1900, Pennsylvania ranked fourth in timber production; by 1920, nineteenth.[13] Clear-cutting of the hemlock and mixed hardwood forests left woodlands depleted, and most of the lumber companies left Pennsylvania for the Great Lakes region.

In their wake, they left denuded hillsides and fields of stumps. Where once deer had found food and shelter in the hemlock forests, the cutover districts provided neither. Without forest cover, deer were more visible to hunters, more vulnerable to the elements, and less likely to breed.[14]

And so populations plummeted. According to Henry Shoemaker, Pennsylvania hunters killed only 150 deer in 1898. To be sure, Shoemaker did not reveal how he arrived at his tally. It seems likely that hunters took many more deer that escaped the notice of whoever was counting. But the number suggests how far deer populations had fallen in the late nineteenth century.[15]

This decline in game populations precipitated widespread concern, particularly among sport hunters like John Phillips. Responding to their demands, the state legislature created the Pennsylvania Board of Game Commissioners in 1895, directing them to secure increases in game populations. The Game Commission did not begin by trying to restore habitat, however. Instead, they restricted hunting by closing seasons for deer and other game, and outlawing many hunting practices, such as jack lighting and hounding. A chief target of conservationists was the game market, which they blamed for much of the decline in wild animal populations. By the turn of the century, it was illegal to buy or sell the meat or hides of many game animals, including deer. In an effort to eliminate interstate traffic in game, the shipment of any animals beyond state boundaries was banned.[16]

As conservationists attempted to regulate the hunting community, the ecosystems of the cutover districts continued to change. Ecosystems respond, sometimes rapidly, to shifts in their composition. In the absence of forest canopy, the sun shone on bare ground that had been shaded for many years. By the early twentieth century, secondary succession was widespread on many of the cutover districts. Among the hemlock stumps, blackberries and other sun-tolerant species soon predominated.[17]

Although they can survive in forests at almost any stage of development, white-tailed deer thrive best in two types of ecosystems. One is the mature forest, in which trees have thinned in places, allowing sunlight to penetrate the canopy and generate shrub growth. The canopy provides needed cover, and the shrubs supply food. But deer also obtain vast quantities of nutrients from forest stands in early stages of plant succession, before the tree canopy is fully developed.[18] Secondary succession proceeded quickly on the old hemlock forests, and within two decades the cutover districts had become prime wildlife habitat. In 1907, the *Harrisburg Courier* observed, "The cutting down of the original open forest has been succeeded in many localities by a dense second growth, which has escaped burning, and sometimes by a third growth difficult to penetrate."[19]

As the Game Commission continued to formulate a policy to increase deer numbers, the laws seemed increasingly redundant to local residents in the cutover districts. In 1907, the state banned the hunting of does and young deer with the passage of the Buck Law.[20] "The State legislature passed a law prohibiting the killing of does," wrote Clarency Seely of Tioga County (part of the state's rugged mountain territory in the north) in 1908, "and I suppose some think it will just be the saving of the deer." Seely went on to describe the changing landscape which, to his mind at least, was more responsible for increased deer numbers: "All the state could or ever would do in the way of legislation is not a drop in the bucket compared to the protection the deer have from the blackberry briars. The blackberry briars were the only salvation the deer had. The thousands of acres that have had the timber taken off is nothing but brush almost impossible to get through."[21]

The new growth also provided forage and shelter for other animals. In 1906 the Game Commission remarked that nongame birds "were far more numerous last summer [than] they had been for a number of years," and that fruit and berry farmers were complaining of excessive crop damage. By 1911, the commission reported rabbits "in excessive numbers in many sections of the state"; in some areas they were "really becoming a menace to growing crops and trees."[22] In the "rough mountain sections" there was "a most marked increase in the number of bears."[23]

Local residents would no doubt have objected to the state's assumption of control over hunting in any case. But the changing natural environment aggravated the dispute between local people and state authorities over how best to interact with that environment. A sudden increase in bear populations coincided with the first closed season on bears in 1905, a perplexing situation for locals. In 1907, the *Harrisburg Courier* described the bear during the closed season as "a chartered libertine, protected by the State and free to come out of his sheltering bush and eat and destroy what he pleases without being shot. This the farmers of Tioga [County] cry out against." Farmers complained of bear raids on honey and ham stores, as well as loss of stock. Bears "occasionally on a Summer evening take possession of the front porch and compel the farmer and his family to take refuge within the house." The sympathetic journalist summed up the view of many local residents: "If it is desirable to preserve and multiply the black bear in Pennsylvania, it is still a question, does not the dense second growth of timber on the mountain sides offer him sufficient protection without sheltering him in his raids on farmers by the penalties of the game laws[?]"[24]

Although there was some violation of the Buck Law, especially in the early years after its passage, it was effective in decreasing the hunting of female deer. And, just as heavy timber harvest had altered ecosystem dynamics, so too did the creation of new hunting patterns, in which hunters no longer killed female deer. By the turn of the century, predators such as the puma and wolf had been eradicated in Pennsylvania. The abundant habitat and absence of human or other predation on female deer meant that the population grew, as hunter harvesting of male deer alone cannot control the size of the population.[25] By 1911 the Game Commission was reporting the expansion of deer populations into "territory where no deer have been found in a wild state for many years."[26]

At the same time that deer were increasing, conservationists were begin-ning a final push to eliminate the game market and market hunting within the state. Although there had been laws against market hunting since at least the 1890s, sales of venison and other game products continued in some areas until well after 1900. Eliminating the last vestiges of the cash-for-game system would entail construction of a new, rigidly controlled market to challenge it. Because market hunters could turn large numbers of wild animals into cash, they were renowned for profligate killing. To conservationists, the elimination of market hunting was sure to create an abundant wildlife population for the public to hunt.

Pennsylvania conservationists had succeeded in outlawing the sale of some game birds, including ruffed grouse and quail, by the early 1910s. But it was still legal to sell others, such as ducks. Shipping grouse and quail in boxes labeled "ducks" became a standard practice of illicit market hunters. Butchers and game dealers in Pittsburgh, Philadelphia, and Harrisburg were eager to obtain grouse for urban consumers, and they paid rural market hunters moder-ate but attractive prices for their hunting booty. The sportsmen's journal of Pennsylvania, *In the Open,* editorialized about the threat of this market system in 1912: "It is evident that so long as city dealers hold out temptation to the poor mountaineers and country boys in the shape of alluring prices for grouse and quail, that the birds will be both shot and trapped in violation of law, and that the only way to stop the traffic is to have sufficient wardens to watch for illegal shipments and sales of game."[27] Indeed, the Pennsylvania game markets were centers of commercial networks that reached far into the surrounding countryside. State authorities reported that when adjoining states banned the sale and export of game, market hunters in those states covertly shipped game to Pennsylvania contacts, who then sold the game in Pennsylvania cities.[28]

This thriving underground market also allowed Pennsylvania's own market hunters to trade game. In particular, residents of remote areas frequently turned to illegal market hunting as a way of securing scarce cash. Because the northern mountain counties were vast, rugged, and thinly populated, effective patrolling was almost impossible. Violations of state game law were commonplace. For many urban sportsmen, this situation was unacceptable. Editors of sportsmen's publications frequently denounced the lawlessness of remote regions. An edi-torial in *In the Open* stated: "The law is being violated every day in the moun-tain sections and will continue to be violated until such time as the moun-taineers are made to feel its power."[29] Market hunting and illegal game sales were part of an environmental crisis, "slowly but surely aiding natural conditions and an army of hunters in exterminating the native game birds of the United States."[30]

The remoteness of "mountaineer" towns from state authorities meant that local control often prevailed over nearby hunting grounds, and that deer were often treated as a local commons despite state law. A tradition that had been around for decades—or even centuries—in many areas, trading game for cash was still a respected way of making a living for many locals. In painting rural people as victims of fast-talking, wheeling-and-dealing urban game purveyors, conservationists obscured a vital relation between urban markets and life in the mountains. Game marketers were in part reaching out to rural hunters, "tempting" them to sell birds in violation of the law. But rural, lower-class hunters were full participants in the exchange. They made the choice to kill the animals and offer them for sale. Selling game in violation of state law was proof that despite de jure construction of a state wildlife commons in the game code, the de facto existence of the local commons continued. As much as it was the domain of urban "temptation," the game market was an arena for the exercise of local prerogative and power.

The contest between local commoners and advocates of the state was characterized by class divisions, nowhere more visible than in the struggle over market hunting. Those who demanded more wardens to patrol the mountains were mostly urban sportsmen. Landowners, particularly those in better farming areas, also supported rigorous law enforcement as a way of protecting their property. At a time when there were only ten paid game wardens in the state, the services of law enforcement officials frequently went to farmers who had personal connections to authorities. As *In the Open* reported in 1911, "The trail of destruction, not only of game, but of chickens and farm animals, left by the out-of-season shooters, created a wave of resentment among the farmers this year, and they were quick to appeal to friends and acquaintances in the cities, to have paid game protectors sent to their assistance."[31]

By 1912, sportsmen were courting these rural landowners, hoping for their support in the campaign to end market hunting. But before market hunters could be pushed from the fields and forests, the state would need better law enforcement. Shutting down the market in game would require full-time, salaried game wardens, as town constables seldom enforced divisive game laws. Hiring game wardens cost money, and the only way to garner enough money was to sell hunting licenses. Farmers and market hunters both had opposed previous attempts at passing a resident hunter's license law. As the editors of *In the Open* noted in 1911, "It was the apathy of the farmers and the open opposition of the all-year hunting fraternity in the mountains that caused the defeat of the last Hunter's License bill, and the consequent smallness of the force of real, efficient game protectors."[32]

Again and again, Pennsylvania's urban recreational hunters pointed out that the worst game law violations and the most persistent market hunters were to be found in states where resident hunters did not have to buy licenses. License sales in many states provided funds to employ full-time game wardens who shut down a great deal of illegal hunting activity. By 1911, thirty-four states had laws requiring resident hunters to buy licenses. Without a fund from sale of resident hunting licenses, Pennsylvania had no way to pay for "an adequate force of paid officers to effectually suppress the sale of game." As a result, Pennsylvania gained notoriety as a "fence" state, a buyer of illegal goods for an interstate network of illegal game dealers. When federal agents and Pennsylvania authorities raided several illegal game dealers in the state, they reported that "every single bird killed, shipped and sold came from states where there was no resident Hunter's License law."[33]

Given that the license law was to be a tool for suppressing the game market, it is easy to overlook the irony that it represented a hybrid of market dynamics and state power. This was especially apparent after the passage of the Resident Hunter's License Law in 1912. Thereafter, all Pennsylvania hunters had to buy a one-dollar hunting license before taking to the fields each year. In a concession which doubtless did much to secure their support for the bill, landowners were allowed to hunt their own lands without buying a license.

Although the state estimated that 100,000 landowners would hunt their own estates without licenses, 300,000 resident hunters bought licenses in the first year of license sales.[34] The result was a virtual revolution in funding for game conservation. Annual legislative allotments for game conservation had grown steadily, from $400 in 1897 to $49,000 in 1913, but even the most generous funding from the assembly did not approach the money received from the sale of licenses. License sales for 1913 alone gave the commission almost $300,000, a 500 percent increase in funding.[35]

Although hunting licenses on one level are a constituency tax, another way of looking at them is to consider them as bills of sale in a new market, one in which the state rigorously controls all transactions and thereby accrues tremendous fiscal and administrative power. The Resident Hunter's License Law was the foundation of a new system of exchange. Once, markets in venison and deer hides had been predominant features of Pennsylvania life. Now, instead of selling meat or hides, the hunt itself was for sale. Each person who wanted to go hunting went first to state authorities and paid one dollar. The license he received in exchange certified that he had bought "a hunt." Buying a hunt did not guarantee the hunter a deer or any other animal. Strictly speaking, the game was not for sale; the opportunity to walk the fields with a gun (and fire it at a deer) was. The commodity was no longer the deer, but the activity. In addition, even that exchange was subject to many conditions. Any violation of game laws, such as killing a doe or hunting out of season, allowed authorities to confiscate the carcass and seize the hunter's license, in effect nullifying the sale of the hunt.[36]

At stake in the struggle over market hunting were rival visions of the countryside, and rival notions of how best to secure abundance from it. On the one hand, local market hunters and their urban game dealer allies sought to direct wildlife to local use. Local prerogative brought game to market channels and provided local hunters and game dealers alike with cash. Localism was intimately tied to entrepreneurialism, a system whereby any local person could establish a business enterprise by investing in a gun and ammunition, assuming the risk for failure if he or she could not kill any marketable animals. Abundance on the local commons came not so much from plentiful game as from the ability of local people to exchange what game there was for cash.

On the other hand, urban recreational hunters and their rural landowner allies saw only dearth in the local commons. To their way of thinking, the local commons and market hunting meant a landscape devoid of wildlife, criss-crossed by "lawless" hunters who shot stock and damaged other farm property. The conservationist alliance was not without internal divisions: farmers supported game laws to protect property; sportsmen supported them to protect deer and other game. But together, farmers and sportsmen formed the backbone of conservationism, and they extolled a vision of abundance in which there would be plentiful wildlife for recreational hunters and well-protected property for farmers, all underpinned by the state sale of hunting privileges. The battle over market hunting was not over whether markets should be attached to the game, but what should be for sale.[37]

The shifting dimension of market exchange—from cash for game to cash for hunt—was coeval with another shift, this one in the scope of state power. In the days of market hunting, any person who acquired game could sell it. Now, the entrepreneurial character of hunting abruptly disappeared. Many industries arose around recreational hunting, among them guiding, outfitting, and taxidermy. But the core market exchange, cash for a hunt, was strictly controlled by the state of Pennsylvania. No other agency, individual or corporate, could sell access to the hunting grounds the way the state Game Commission did. A landowner could hunt his own lands without a license, and farmers could still turn game to cash, indirectly, by charging "trespass fees" to visiting hunters. But a primary requirement of such transactions was that the hunter who bought the right to stalk the owner's land also buy permission to hunt the public's deer, paying the state for a hunting license. Through the Resident Hunter's License Law, the state established a monopoly over almost all legal hunting.

This market did not function like other monopolies, for state power to set prices for the hunt was circumscribed by the willingness of constituents to pay them. In a free market, a monopolist has the power to increase prices of the monopolized commodity without fear of competition. Here, the state could not raise prices of hunting licenses without fear of political retribution in the legislature. The balance of political power thus shaped the market economy of the privately owned, state-managed hunting grounds.

The market in hunting privileges initiated a host of political changes, and insofar as it helped effect changes in the way people interacted with the forests, it also contributed to ecological shifts. The interface of politics and ecology was perhaps most apparent in the complicated interplay among license sales, deer policy, and deer habitat. Deer were the most popular game species. The state did not sell deer, but the best way to sell a large number of deer licenses would be for a multitude of individual hunters to believe that they stood a reasonable chance of killing a deer in a given year. Early on, the department noticed that there was a direct and positive correlation between deer populations and the number of licenses sold. Joseph Kalbfus commented on the phenomenon in 1916: "Because of the fact that deer hunters were so successful last year, many men who never saw a deer in the woods went to hunt these animals during the season just closed."[38] The more deer there were, the more people bought hunting licenses, the more cash the department had on hand.

Along with a greater impetus to manage for plentiful deer, the Resident Hunter's License Law made the Game Commission more dependent than ever on urban sport hunters. Sportsmen had always been the chief constituency of the game department. But landowners and farmers had been strong supporters of state game authorities owing to their desire to control hunters on their lands.

Now recreational hunters provided state authorities with almost all their finances; some farmers might buy licenses, but they had no direct connection to state authorities on the order of the sportsmen's. Indeed, because they could hunt on their own lands without licenses, there was a strong impetus for farmers not to contribute money to the Game Commission and thereby not to buy themselves a voice in the formulation of game policy. The Game Commission became more than ever a sportsmen's advocate. Joseph Kalbfus remarked in 1915 that the significance of the Resident Hunter's License Law was "that it places the sportsmen of this Commonwealth in a position where they, and they alone, are paying for protection to the birds they desire to kill as game birds, and I feel that because of this condition they alone, the sportsmen, should be considered in the enactment of laws relating to the killing of game, to the length of seasons during which birds may be killed, and to any other conditions relating to the taking of game."[39] So long as the sportsmen who paid the license fees and farmers who owned deer habitat were united, the Game Commission's task was relatively simple. Once these two groups fell to fighting, the new order of the hunting grounds was seriously jeopardized.

Early on, few had any inkling of trouble. Funds from license sales brought a sudden growth in the administrative power of the Game Commission and freedom from tax-based legislative allotments.[40] State authorities hired more game wardens and moved against illegal market hunting operations with new confidence. By the mid-1910s, authorities had deemed the anti–market hunting campaign a success.[41]

Game Commission power was at an unprecedented high, and authorities struck out in new directions. Funds from license sales gave them the ability to lease lands for state game refuges, and in 1919 the commission began buying lands for public hunting grounds. The commission's first land purchase was a parcel in the newly abundant deer habitat of Elk County. Since 1906, Joseph Kalbfus had been importing deer from northern Michigan, as well as other game; plentiful funds allowed him to pursue such programs on a larger scale.[42] With a lucrative business in the sale of hunting privileges, the state had wider latitude than ever in its campaign to create an abundant wildlife resource under centralized control.

But this freedom soon faced new challenges from an unexpected source: the sportsmen's erstwhile allies, the farmers. As authorities extended their political power, rededicating themselves to increasing game populations, deer continued to expand their range. Inevitably, they spread into farm fields.[43] And, as local deer herds increased, so too did farmers' complaints of crop and property damage. The Game Commission was not completely unsympathetic. Joseph Kalbfus noted in his report of 1915: "Orchards, together with fields of growing oats, and buckwheat, have undoubtedly been seriously injured by deer in this State during the past year . . . and it is not fair or just that the owners of these growing crops should be compelled to have their crops destroyed without the hope of a return."[44]

But more often, Game Commission leaders were reluctant to accept the complaints of farmers, and frequently considered them opportunistic. With huge funds at its disposal, the Game Commission was increasingly the target of farmers who demanded compensation for deer damage. By the mid-1910s, as Kalbfus reported to the Game Commission, "deer damage" was becoming a euphemism for winter-killed plants, the ravages of domestic stock, and failed crops in general, inspiring his 1915 remark that "all farming in the State is not limited to a cultivation of the soil."[45] No doubt some of the disagreements between game authorities and local farmers were honest ones, but specious claims were widespread. Appealing for state compensation in such cases constituted a covert resistance to state authority in wildlife, a way of making the new, sportsmen-dominated state commons function to the benefit of local farmers.

There was little doubt that deer populations were rising quickly. We may impute the magnitude of the increase from the Game Commission's figures for the total legal deer take. In 1913, some 300,000 licensed hunters took 800 bucks out of Pennsylvania's hunting grounds. In 1915, when only 250,000 licenses were sold, the kill of bucks totaled more than 1,200. By 1919, when total license sales had increased 30 percent over 1913 figures, to 400,000, total buck kill had increased more than 250 percent, to 2,939.[46]

Had the increase remained steady, it might have been easier to adopt new hunting policies to respond to it. But the forest ecosystems of Pennsylvania continued to evolve, frustrating attempts to adapt resource uses to the land. By the 1920s, areas of the forest that had been clear-cut in the second half of the nineteenth century and brush-covered in the first two decades of the twentieth century were at an intermediate stage of development: the so-called pole stage, when trees resemble poles thirty feet high or taller. In an old-growth forest, there would have been plants and trees of varying ages, thus guaranteeing that some would be within reach of the deer, while others would be old enough to die and clear a hole in the forest canopy. The patch of sunlight falling through the gap would allow shrubs to grow on the forest floor, providing more deer food. But when succession species mature on old clear-cuts, they are uniform in age and height, and few die to leave sunlit clearings where deer forage can grow. The uniform age of tree stands meant that eventually most potential forage would be out of the deer's reach.

This was the situation in Pennsylvania by the 1920s. Oak, chestnut, maple, birch, aspen, and pin cherry had succeeded on the old white pine and hemlock forests. Now they were growing tall enough to escape hungry deer; shade cast by their leaf canopy meant the end of the lush carpet of shade-intolerant plants that had provided nutrients to the expanding deer population. There were more deer every year, and each year the forests produced less food for them.[47]

In the twenties, the state commons began a subtle but ominous shift. There were still many deer, but with little to eat in the forests, they looked elsewhere for sustenance. Increases in deer populations ensured increased crop damage even when wild forage was available, and when forests became depleted, crop damage intensified.[48] In some places the restocking efforts, the Buck Law, and forest succession were creating a radically new ecology, in which the state commons—the deer—intruded on private farms as never before. The Game Commission noted this phenomenon in 1924: "The situation with reference to deer is becoming somewhat serious in certain agricultural sections. Female deer have increased so rapidly in sections stocked early in the history of the Board and are encroaching upon surrounding farms to an extent that is disconcerting, to say the least. This condition has been aggravated through the protection of forests against fires, with the result that undergrowth has been choked out in certain forest areas and the deer are compelled to depend upon surrounding territory for suitable forage."[49] As changing ecosystems drove deer to farm crops, they in turn drove farmers to greater conflict with the Game Commission.

To ameliorate the deer food problem, state game authorities attempted to reestablish control over the public's deer. They began a campaign of planting vegetation for deer to eat in some areas, and trapping animals from overpopulated districts for transportation to others.[50] But trapping was expensive and ineffective, and hungry deer ate the planted seedlings before they could mature.[51]

Attempts to protect farmers from deer were no more successful. In 1923 the state legislature passed the Deer-Proof Fence Law, which provided farmers in deer-ridden areas with subsidies for deer-proof fences. The commission was bewildered by farmers' reluctance to exploit this benefit, complaining that "comparatively few farmers feel they can afford to pay their portion of a suitable fence, even though their share would not exceed the cost of an average woven wire fence."[52] Subsequent amendments did little good: in 1925, the law was changed so that the Game Commission provided wire and staples, farmers being left to provide posts and labor. Still, few farmers used these measures to save their crops.[53] Fencing was a limited option, anyway. Whether it was a motivating factor in their decisions not to fence their fields, the community relations dilemma that fences created for farmers was a sign of how much the changing commons affected local life. Farmers soon noticed that deer, fenced out of one field, moved to unfenced fields in greater numbers than ever.[54] The farmer who fenced his fields against deer thereby turned more of them on his neighbors.

As early as the late 1910s, deer population dynamics had many observers questioning the viability of the new hunting order. To some, the notion of proper hunting behavior seemed inordinately shaped by notions of gender. Conservationists had adopted much of their legal code from sportsmen's ideas about the sanctity of femininity and young life, ideas which inscribed middle- and upper-class notions of gender on the natural world; bans on the killing of female or immature deer were as much ethical prescriptions as they were a means of manipulating nature to produce abundant deer. Perhaps rural people never adopted such conventions; there had always been some local resistance to the Buck Law. But the expansion of deer herds brought increasing violations, concentrated enough in certain localities for state authorities to suspect a local "plan or agreement" to subvert the Buck Law.[55] By killing female deer, rural locals were not only defying state authority, but challenging sportsmen's cultural attachments to the animals as repositories of motherhood and femininity.

Rural residents who killed does probably would have agreed with those conservationists who had a more scientific appreciation of deer population dynamics. In 1917, Joseph Kalbfus himself lobbied the legislature for a legal doe hunt every five years. The legislature, having taken years to endorse the sportsmen's program that banned doe hunting, was not willing to consider an about-face on the issue of killing female deer. It opted to retain the Buck Law without amendment. Kalbfus was grim about future prospects. Noting that more than 1,700 bucks had been killed the previous season, Kalbfus predicted that his successor would "watch the fur fly about ten years from now." The widespread sentimentality surrounding female deer had been invaluable in building a conservationist consensus among Pennsylvania voters. Now, with deer numbers shooting upward, Kalbfus saw the potential for angry battles over deer policy in which sentimentality would hinder efforts to control the deer herds. As he told a colleague, "We oversold our customers. The volcano of sentiment we built is likely to blow up."[56]

The remark was prescient. Increasingly, divisions appeared between farmers and urban sportsmen on the issue of hunting female deer. Many farmers began resisting the state's sanctions against doe hunting, and some angry farmers challenged the Buck Law in court. In a 1916 appeal of a doe-shooting conviction, the defendant's attorney argued that the state, as the owner of the deer, should take responsibility for deer damage or allow farmers to shoot the animals. "If a farmer's crops are destroyed by his neighbor's cow, he can recover from the owner of the cow the value of the crops destroyed. The state of Pennsylvania, the owner of the wild deer who destroy a farmer's crops, cannot under our law be made to pay the damage. The only remedy the farmer has is the extermination of the deer."[57] In 1923, the defense of a man charged with shooting a doe in his employer's fields centered on the difficulty local residents were having with deer. "A large number of wild deer have been for years living in this mountain," explained the defense lawyer. "They not only damage the fruit but the vegetables and the gardens and the crops in the fields as well. They have rendered it almost impossible to have a vegetable garden. They have driven fruit farmers from their farms and, if allowed to continue their maraudings, these farmers, who have made large investments and who year after year spend much time and labor in caring for the fruit trees, will be compelled to abandon them; and there will be, instead of blossoms and ripe fruit found in abundance, thistles, thorns and briars."[58] It is probably true that the attorney overdramatized this picture in an attempt to secure an acquittal for his client (which he did). But the imagery revealed the increasing divergence between urban and rural priorities in the state wildlife commons. Thistles, thorns, briars, and deer were emblematic of wild nature. The "wild" that urban conservationists aspired to create, a landscape of abundant deer, was consuming the domesticated farmer's world of verdant gardens and abundant orchards. To defend against this onslaught of nature, rural people began to argue in increasing numbers that female deer must be killed. The dispute over proper responses to the changing commons hung in part on questions of gendered behavior, on how to define manliness in the hunting grounds.

From the rural vantage point, conservation seemed to have turned the countryside on its head. For those farmers who supported it in earlier days, conservation had been a way of promoting a particular social order in the countryside, an order in which the state lent its power to defending their property against invading hunters. Suddenly, the hunting grounds seemed strangely disordered, not by social change, but by a peculiar shift in which deer, rather than hunters, were the invaders. In seeking an appropriate response, locals now found themselves hamstrung by the state authorities who had been their allies. Rural people's preferred measures for dealing with the problem, such as killing female deer, were blocked by a remote Game Commission and urban sportsmen.

When a commons changes, it might seem a matter of course that its users change the terms of access to it. But attempts to alter legal conditions of access to deer spawned battles that pitted the countryside against the city with increasing frequency. In 1923, in the vicinity of Gettysburg, farmers reported being "overrun with deer." State authorities, agreeing that there were far too many deer in the area, arranged for a special doe hunt to reduce deer numbers. Sportsmen immediately objected on the grounds that killing does was unethical. Plastering the area with yellow signs reading "Don't Be Yellow and Kill a Doe!" they dissuaded many would-be hunters from taking part. One hundred special doe permits were issued, but only eight does were taken.[59]

Other attempts at doe hunts met with a similar fate, illustrating how, all too often, recreational hunters placed a different value on deer than did their farming contemporaries. Indeed, the dispute hinged on competing perceptions of female deer and involved constructs of gender. To a considerable degree, manliness for urban sportsmen served urban ends, the demands of work and home in the city. Urban sportsmen's idealization of female deer in the fields and forests expressed, in part, their notions of women's place—particularly upper-class women's place—in urban settings. Doe and woman were both reified as mother, noncombatant and, by extension, noncompetitor in the male world.

The identity of rural manhood might incorporate some of these ideas, but in a significant way it was usually attached to the farm field; being a successful man in farm country meant raising a large and profitable crop. For rural men, female deer threatened manly identity by endangering the harvest. As the deer population in the country increased, so too did clashes between city and country over what the deer irruption meant and how to respond to it as hunters and as men.

The state did accommodate farmers in some ways. Since game laws first appeared in Pennsylvania in the late nineteenth century, the right of farmers to kill animals causing property damage had been protected.[60] But where deer were concerned, that right had long been constricted by other regulations. Farmers who killed deer to protect crops had to notify a game warden within twenty-four hours, and the carcass had to be dressed for donation to a hospital or other charity. As deer invaded farm fields in increasing numbers in the twenties, farmer complaints about the restrictions succeeded in loosening them. In 1925, farmers in areas particularly hard hit by deer received the right to keep carcasses for food.[61] Between 1926 and 1928, farmers reported more than 1,200 deer killed in defense of property.[62] In 1927, the state Game Commission hired local men in two counties to kill deer in areas where deer damage was pronounced. Although they killed almost two hundred animals, the effect on deer populations was negligible.[63] In 1929, the legislature extended to farmers throughout the state the right to keep deer carcasses for food.[64]

And still the commons continued to change in ominous ways. Not only were there numerous deer, but by the late 1920s, many were starving to death amid ravaged farm fields. State game authorities commissioned Vernon Bailey, chief field naturalist for the U.S. Bureau of Biological Survey, to examine deer range and assess its condition in four Pennsylvania counties. Bailey reported that there were so many deer in Elk, Clearfield, Mifflin, and Center counties that there was not enough food for them. "From life-long observation on the food habits of deer over practically all of the United States, I am aware of their varied food habits but never before have I found them in one locality feeding on so many different kinds of vegetation."[65] Estimating the deer population of the state at more than one million, Bailey called attention to the "deer line" in the forests and the fact that its height—six feet in some places—kept the young fawns from eating at all and hastened their starvation. Deer overgrazing was destroying the habitat for other game species, notably rabbits and ruffed grouse.[66] His recommendation: to begin the systematic hunting of does again, thus bringing the herds back into balance with available food and cover.[67]

The termination of doe hunting had effectively introduced a new factor in the ecology of deer habitat. With proportionally fewer female deer being taken from the ecosystems of Pennsylvania, the reproductive capacity of the deer herds had increased dramatically. Forest ecosystems were responding in ways that ultimately diminished the herds: with forage unavailable, does would become less fertile, and fawns would starve in large numbers. Eventually adult deer might also starve, and the forests would then recover their ability to generate forage.[68]

But for many, deer were much more than ecosystemic components; to conservationists and their opponents, starving deer were a sign not of an ecosystem righting itself but of crisis. Recreational hunters in particular felt obliged to save deer from famine, and the Game Commission disseminated a number of pamphlets on providing winter food for deer and other game species.[69] It is unlikely that these nutrients were sufficient to affect the population, but the gesture itself illustrates the central paradox in conservationist values: having acted to allow so many deer to live as a means of making the woods "wild," the sportsmen now felt it their responsibility to intervene in natural processes and save starving deer, essentially domesticating the nature that deer represented.

The failure of these and other measures to establish control over the unpredictable state commons was manifest in numerous ways by the late 1920s. In Elk, Clearfield, Center, and Mifflin counties, sportsmen estimated that more than 1,000 fawns died from starvation in February and March 1928. One deputy game protector buried more than 200 dead fawns. So many of the animals were dying along stream courses that biologists expressed concern for the safety of the water supply. More than ever, sportsmen were complaining about stunted deer, poor antler size, and rotten carcasses of starved fawns along stream banks in the spring. In 1928, the state held its first statewide doe hunt.[70]

The hunt did not go unopposed. Some sportsmen's groups filed suit against the state to have the hunt stopped, and the state attorney general found that the Game Commission did not have authority to declare a doe hunt. Through several innovative measures, the commission was able to declare an "antlerless" hunt in fifty-four of the state's sixty-seven counties. Sportsmen maneuvered to have many of these counties closed to the hunt. But ultimately the vociferous objections of recreational hunters no longer had the appeal they once had. In contrast to the 1923 doe hunt, when only a few does were taken over the objections of sportsmen, in the 1928 hunt more than 25,000 "antlerless" deer were taken from the state commons.[71] Disputes over when to declare doe hunts would be commonplace long after the 1920s. But after 1928, hunting female and immature deer under state control was widely considered an occasional, if distasteful, necessity.

In the earlier days of the campaigns against immigrant hunters and market hunting, urban sportsmen and rural landowners had combined to bring state authority to the rural hunting grounds. But this alliance of urban and rural elites was increasingly strained by the changing commons of the 1910s and 1920s. Acquiring political power through the manipulation of markets in hunting was relatively easy compared to exerting real power over local ecosystems. Pennsylvania's forests were themselves responding to earlier markets in timber harvesting and shifting uses of the forest, including the suppression of doe hunting. Social conflicts over how to use those forests were divisive and bitter; when the commons changed, political and social alliances among the commoners also changed.

Despite their differences, sportsmen and farmers shared a reliance on ecological dynamics to provide sought-after products of local ecosystems. Conservationists wanted to bring wild animals from the land, while farmers tried to cultivate fruit, vegetables, and livestock. Both relied on natural processes of photosynthesis and growth. The crucial difference was that farmers required fixed crops with predictable natural cycles and controlled environments free of pests, whereas conservationists relied on more autonomous natural processes that they understood in only the most rudimentary way. Deer were less controllable, more "wild" than onions or cows, and therefore they were less predictable. To farmers and others whose livelihood was threatened by the surging deer population in the countryside, the reordering of human relations with the wild had created a strangely disordered environment, in which a deer's livelihood occasionally seemed more protected than their own.

But if deer as "wildlife" represented a baneful wilderness to farmers, in fact the landscape of abundant deer in the 1920s were no less a human creation than fields of corn or wheat. Conservationists did not need to labor to bring forth deer the way farmers did to raise crops, but secondary succession in the cutover districts and growth in the deer population were ecosystem responses to logging and the ban on hunting does. They were natural responses to human activity, similar in that sense to the germination of cash crops on a farmer's plot. Farmers sought to change the land to direct its produce to market; in a sense, so did conservationists, who helped promote abundant deer habitat as a way of stimulating demand for it as a hunting ground. Attorneys for farmers could paint a picture of their clients defending domesticated landscapes from an advancing frontier of "wild" deer, thistles, thorns, and briars. But even here the choice of metaphors reflected a deeper irony. Thistles, thorns, and briars are succession species, characteristic of landscapes disturbed by burning, plowing, or other human activity. In this sense, they were apt symbols for the advancing deer herds, which were a product of human manipulation of the forests.

At issue in the battle between sportsmen and farmers were not the merits of wilderness or the benefits of civilization. This was a competition between markets in produce and markets in recreational hunting, between cultural ideas of wildlife that held that deer were a threat to livelihood on the one hand and a fragile symbol of nature on the other, and between the power of local people to define their own interactions with the land and the power of the state to restrict those interactions. All these conflicts surfaced repeatedly in the dispute over how people should live amid the state commons, the ever-changing and unpredictable deer herds.

By the 1920s, rural Pennsylvanians could consider stories like that of John Phillips—who feared that he had killed the last deer in Pennsylvania—nostalgic and naive. The lesson of the 1920s was that ecological changes rendered untenable certain assumptions once shared between sportsmen and landowners, who together had been architects of the state commons. At one time, it was thought that plentiful game and orderly farm fields went hand in hand. Few believed so now. The conservationist coalition fell apart as abundance for urban sportsmen—the deer—threatened or destroyed the abundance of farm life—the crops.

In the conflicts between the interests of city and countryside, it is important to recall that neither side was unified. Class divisions characterized city and country, and class lines that divided rural Pennsylvanians ensured that the state commons in deer meant different things to different rural people. For farmers on better lands, the battles of the 1920s and subsequent decades required lobbying the legislature, negotiating with the Game Commission, and at times defending themselves in court on charges of doe shooting or other measures taken in defense of their farms. But for people living in more remote areas—the impoverished "mountaineers and country boys" in the northern counties—the ascendance of the state commons meant a greater threat to livelihood. For these people, hunting had long supplemented a meager living. Owners of lands of marginal value for farming, people of the northern counties were known as "hill farmers" for their practice of eking out crops in this poor land of steep mountains. In 1950, the Game Commission's Roger Latham summed up a half century of deer conservation battles. For decades, he noted, there had been a pronounced split between the perceptions of urban sportsmen and those of rural people over how many deer there were and what needed to be done about them. Urban sportsmen frequently did not understand that deer "have learned to dig potatoes, to tear corn shocks apart, and to eat nearly anything that can be grown in a garden with the possible exception of onions."[72]

Farmers were often hunters, and they did not want deer eradicated from the forests any more than sportsmen did. But besieged by thousands of the hungry animals, farm living was substantially threatened. Particularly hard hit were farmers on poor land: "At best a farmer can hope to make no more than a meager living from marginal land—and when his entire buckwheat crop may be lost, much of his winter's supply of potatoes dug out and eaten, and his truck patch ruined, this loss may mean the difference between success and failure. Many 'hill farmers' have been driven from their land because they could no longer compete successfully with the deer and make a living for their families."[73] As Latham observed these sad events in 1950, J. Q. Creveling, an elderly Pennsylvania sportsman, was composing his own summary of conservation struggles. Creveling saluted the successes of conservation, recalling the blasted game herds of the turn of the century. "In the year 1900," he recalled, "while on a fishing trip on the Willowamac Creek in New York, I had run across a fresh deer track in the mud along the creek." Although he was a dedicated outdoorsman, deer populations in 1900 were so low that Creveling had never seen one in the wild before. "I followed the track into a swamp, thinking I might be able to see the deer, but it had evidently discovered me and I was unable to see it."[74] To urban sportsmen, Creveling's story could be read alongside the account of the Phillips hunting trip. Together these tales became part of a body of environmental morality tales in which unregulated hunting rendered a landscape barren of its essential, "wild" inhabitants. Farmers often enjoyed hunting, as much if not more than urban sportsmen; it seems unlikely that any would want deer to return to the endangered condition of a half century before. Still, if empty deer licks and gameless meadows were both a distant memory and the sportsman's darkest dream, to the hill farmers and others standing armed guard against hungry deer, they could not have seemed such a dismal prospect.

Italian immigrant quarry workers, northwest Pennsylvania, early 1900s.
(Lawrence County Historical Society, New Castle, Pennsylvania)

Rocco Racco at his arrest in 1908. A shadowy figure in Hillsville, he held a grudge against the local game warden according to the testimony of enemies. (Pennsylvania Historical and Museum Commission, Harrisburg)

Deputy game protector Seely Houk shortly before his murder. His campaign against Italian immigrant hunters endeared him to local farmers—and made him many enemies in heavily Italian Hillsville. (Pennsylvania Historical and Museum Commission, Harrisburg)

Early wildlife conservation was often nativist in tone, as shown in "The Regular Armies" of wildlife destruction, produced by early conservationists William Hornaday and Dan Beard. On the right a man bears the sign "We are Aliens. We Kill the Songbirds." The man cradling the shotgun in front is obviously a stereotyped rendering of an Italian immigrant. (William Hornaday, *Our Vanishing Wildlife* [New York, 1913]).

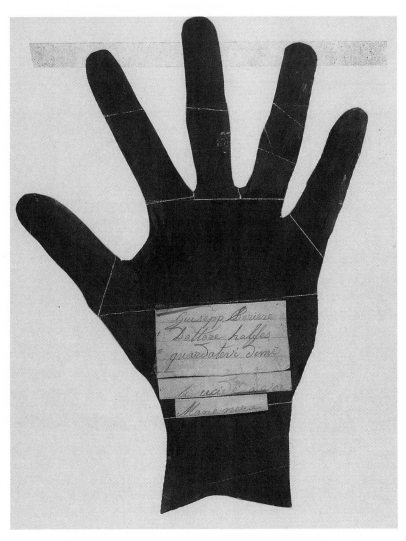

A Black Hand death threat sent to the Game Commission,
addressed to the secretary, Dr. Joseph Kalbfus ("Dottore
Kalfes") and his chief game warden, Joe Berrier ("Giuseppe
Beriere"). Kalbfus received three messages like this, but
probably from disgruntled individuals and not, as many
believed, from a criminal organization.
(Pennsylvania Historical and Museum
Commission, Harrisburg)

Early twentieth-century cartoon decrying bird extinctions at hands of hunters. Although habitat destruction often played a role in these extinctions, lower-class hunters often shouldered most of the blame. (William Hornaday, *Our Vanishing Wildlife*)

SACRED
TO THE MEMORY OF
THE NORTH AMERICAN
GREAT AUK
PALLAS CORMORANT
LABRADOR DUCK
PASSENGER PIGEON
ESKIMO CURLEW
CUBAN TRICOLOR MACAW
GOSSE'S MACAW
GUADALOUPE MACAW
YELLOW WINGED GREEN PARROT
PURPLE GUADALOUPE PARAKEET
CAROLINA PARAKEET
EXTERMINATED
BY CIVILIZED MAN
1840–1910

To many conservationists, market and subsistence hunters were merely "game hogs," whose behavior needed to be constrained or banished from the hunting grounds. (John F. Reiger, *American Sportsmen and the Origins of Conservation* [Norman: University of Oklahoma Press, 1986])

Unidentified Indian hunters under arrest at Magdalena, New Mexico, circa 1910.
(New Mexico Department of Game and Fish, Courtesy New Mexico State Records
Center and Archives, no. 62844)

"An Indian Hunter," according to New Mexico's game warden, circa 1910. The sight
of men like this one made New Mexico's rural ranchers and sport hunters anxious about
wildlife survival and their own control over the countryside.
(New Mexico Department of Game and Fish,
Courtesy New Mexico State Records Center and Archives, no. 62845)

3
"Raiding Devils" and Democratic Freedoms
Indians, Ranchers, and New Mexico Wildlife

In January 1916, the U.S. Forest Service sent a young official named Aldo Leopold on a speaking tour of southern New Mexico. His assignment: to help local sportsmen build cooperative associations to protect wildlife on the national forests. The forester succeeded to a remarkable degree. Large and receptive crowds greeted him at every stop. In the town of Roswell, an enthusiastic audience of three hundred turned out to hear his lecture. "The curious thing," he wrote in a letter to his mother, "is that this thing has been tried time and again right here in New Mexico, and as promptly failed."[1] More victories followed. Two months later, this federal forester led the state's sportsmen into a statewide organization, presiding as secretary for its first convention.

In a region renowned for ambivalence and hostility to the Forest Service, this public enthusiasm for federal authority in the forests was indeed a "curious thing." There were other anomalies here, too. Game conservation was usually a state government activity. Why, then, was the U.S. Forest Service, a federal agency, organizing the state's sportsmen?

On the surface, some of the answers are simple. The Forest Service sought better control over the forests, and championing a cooperative effort in game protection was one way of achieving it. Leopold was particularly enthusiastic about the task, and his congeniality no doubt helped soften some of the anti-federal feelings of his audience. Although he was little known outside the Forest Service in 1916, today his name is practically synonymous with wildlife conservation. He eventually became the nation's first professor of wildlife ecology, at the University of Wisconsin. In 1933 he published *Game Management,* a classic in scientific wildlife protection.[2] His *Sand County Almanac,* posthumously published in 1948, became an inspiration for generations of environmentalists, establishing its author as one of the foremost environmental thinkers of the twentieth century.[3]

For all his later successes, it is this relatively minor accomplishment in

New Mexico that preoccupies us. Leopold's personal enthusiasm for conservation and the larger motivations of the Forest Service aside, there are still unanswered questions about these events that make Leopold's lecture tour as compelling a mystery as any murder on the Mahoning. New Mexico was a patchwork of local commons regimes, with distinctive Indian, Hispano, and Anglo enclaves defining the cultural landscape. Nowhere was local life more varied, nor could centralizing control over the hunting grounds be more challenging. Leopold was a charming man, but personal warmth alone seems insufficient as an explanation for the suddenly close relationship between federal agents and local hunters, many of them small ranchers with a history of hostility to conservation.

Anti-federal sentiment was an inevitable problem for conservationists, especially in the West, where most wildlife was in national forests or other federally owned properties.[4] In New Mexico, as in eastern states, game was in the process of becoming a "state commons." But partly because of the size of federal landholdings and partly because of the scarcity of state funds, here the "state commons" took on a pronounced federal dependence. If the state retained responsibility for making game law, federal officers like Leopold became more and more responsible for enforcing it, and for making federal policy with wide ramifications for "state" wildlife. In due course, national authorities asserted a national interest in the deer, antelope, and other game of New Mexico. Consequently, the state commons came to resemble a qualified "national commons," in which wild animals were the public goods not just of New Mexicans but of the American people. Perhaps the most striking aspect of this development is the extensive local support it received in remote places like southwestern New Mexico. Establishing the reasons for Leopold's popularity in January 1916 contributes to an understanding of the national commons, its painful birth, and its relation to the American West.

There was a particular urgency in Leopold's lectures in these years, a sense that without federal conservation efforts, American wildlife and American hunting were doomed. To him and his sportsmen allies, the ascendance of federal national power was essential to the protection of "traditional" American liberties in hunting—and nowhere were those liberties more endangered than in New Mexico. Since the turn of the century, the state's wealthiest and most powerful interests had steadily privatized New Mexico's wildlife. In other states, landowners could limit access to their acreage, but they did not own the game itself. In New Mexico, owners of enclosed property acquired title to wild animals for a nominal fee. New Mexico's hunting grounds increasingly resembled a European barony in which game belonged to the landowner and others hunted only on sufferance. Leopold's rhetoric was nonconfrontational, unlikely to offend any entrenched political interests, and he never openly challenged the champions of landowner privilege. But it is not too much to characterize his tour and the sportsmen's organizing effort as an attempt to stymie the growth of what Leopold called "the European system" in New Mexico. Their aim was to secure limited hunting rights for middle-class recreational sportsmen; at the same time, their campaign had much darker implications for New Mexico's poorest residents.

For now, a grasp of Leopold's accomplishments in New Mexico requires a look at the system he was attempting to replace. In the American West, where myths of individual freedom and egalitarianism run deep, it seems a particular paradox that European-style game protection dominated a state conservation effort. Thus, two questions guide this inquiry into Leopold's program. First, how did New Mexico's land barons lock up the local commons for their exclusive use? Second, how and at what cost did sport hunters combine with New Mexico's dominant rural interests to save what remained of the "free" hunting grounds for themselves? The story begins in a part of the state far removed from Leopold's first lectures.

The Power of Property

The stronghold of landowner interests in New Mexico game conservation was in the northeastern part of the state, in Colfax County, home of the Maxwell Land Grant. "The grant," as it came to be known, was one of the largest private landholdings in the Western Hemisphere.[5] A brief survey of its background illuminates how its owners came to dominate wildlife conservation, and much else, in New Mexico. The Spanish and Mexican governments that colonized New Mexico in the eighteenth and nineteenth centuries facilitated settlement of the region by granting tracts of land to individuals or communities. In 1841 the president of Mexico allotted lands in northeastern New Mexico to two traders, Guadalupe Miranda and Canadian-born Carlos Beaubien, both citizens of Mexico.

The grant had no clear boundaries, and the upheaval of the mid-nineteenth century allowed a series of owners to expand it substantially. Despite Mexican regulations that limited the grant to 97,000 acres at its inception, the American conquest of New Mexico in 1846 provided a window of opportunity for ambitious land speculators. By then, the grant had become the property of Lucien B. Maxwell, an American trader who settled in New Mexico while it was still a Mexican province. When Maxwell sold the land to an English syndicate in 1870, he claimed that it was "2,000,000 acres, more or less." It passed from the English owners to a Dutch company in the 1870s, and, after a long series of legal battles over who owned what in northeastern New Mexico, the Supreme Court ruled in 1887 that the holding, now called the Maxwell Land Grant, was comprised of 1.7 million acres.[6]

The land grant's confirmation concentrated a great deal of real estate—and tremendous power—in a few hands. The owners of 1.7 million contiguous acres—replete with fine grasses, good timber, abundant minerals, and steadily flowing rivers—would have been influential in any state. But in New Mexico, a mostly arid territory with a small population and little industrial or market infrastructure, their economic and political power was huge. The Dutch owners hired a circle of local businessmen, politicians, and lawyers to ensure the most profitable disposition of the company's immense holdings. Selling land in large tracts became the chief industry of this circle of prominent New Mexicans, who became known as the Santa Fe Ring. Comprised of men who had connections and influence in Washington, D.C., as well as the territorial capital of Santa Fe, and including both Democrats and Republicans, the Ring was tremendously effective at securing its political agenda at national as well as territorial levels. Members of the Ring occupied the governorship almost continually from the late 1860s until 1885.[7] When New Mexico became a state in 1912, politicians, businessmen, and lawyers with attachments to the Ring were still very much at the center of power in Santa Fe.[8]

Among men who bought and sold land in million-acre parcels, game laws might have seemed trivial. But the Santa Fe Ring had more than a passing interest in wildlife. Around the turn of the century, the owners of the grant divided some of its western ranges into a number of very large tracts, amounting to hundreds of thousands of acres each. These they leased or sold to wealthy ranchers. As early as 1901, some of these ranchers on the grant were stocking their streams with fish bought from the federal hatchery in Leadville, Colorado, and by 1903 they were buying bobwhite quail for release on their estates.[9]

Records of the private estates are difficult to come by, but the history of the Adams and Bartlett Cattle Company, among the largest of the grant ranches, suggests how the chief project of the Santa Fe Ring—integration of grant properties into eastern markets—was connected to privatizing local wildlife. In

1900, William Bartlett expressed interest in buying part of the Maxwell Land Grant. Bartlett was a wealthy Chicago grain dealer whose reputation for business acumen was enhanced by rumors that he had once cornered the wheat market on the Chicago Board of Trade.[10] Negotiations between Bartlett and the Maxwell Land Grant Company continued into 1901, when company directors discovered that Bartlett was more interested in turning his estate into a private hunting reserve than a cattle ranch. Looking to close the sale, the company stocked his streams with 50,000 trout from the hatchery at Leadville, and they arranged for Governor Miguel Otero to appoint Bartlett employees as game wardens.[11] Bartlett was satisfied. In 1902, he purchased 300,000 acres from the company. He built three mansions on the estate, one for himself and one for each of his sons. Before long, his streams ran with fish, five of his ranch hands patrolled the property as official game wardens, and he became a seasonal and gregarious occupant of his palatial residence, Casa Grande.[12]

In its usefulness as a draw for Bartlett's dollars, wildlife in northeastern New Mexico was an important part of the emerging linkage between western land and eastern capital, a connection that in many ways reordered local land and other resources to suit the demands of extra-local investors. Game laws reinforced that reordering, giving ranchers like Bartlett vigorous support. In 1901, the territorial legislature outlawed the hunting of deer, elk, mountain sheep, and antelope, "except upon private ground or property by the owner or lessee thereof, or with his permission."[13] As the vast majority of New Mexico lands were either public or in holdings too small to provide big game habitat, the law excluded most people from hunting. Essentially, the only hunters allowed to pursue large game in New Mexico were landowners like Bartlett, large lessees, and their associates. Given their interest in stocking streams, it is likely that grant interests also had something to do with 1903 legislation that provided for the protection and propagation of fish by the owner or lessee of enclosed lands.[14]

In 1905, the Ring's influence in conservation increased, as the legislature adopted legal protection for private hunting reserves like Bartlett's. Laws of that year extended trespass protection to any "owner or lessee" of "any inclosure or pasture in the Territory of New Mexico" who wished "to protect or propagate game birds, animals, or fish within said inclosure."[15] According to the state game warden's report of 1908–9, support for game conservation was strongest "in the principal game regions, but more especially among the large land owners, ranchmen, and sportsmen."[16] Colfax County, the home of the Maxwell

Land Grant, ranked first in total number of game law arrests in 1908–9, with twenty.[17] By the time Aldo Leopold arrived in New Mexico in 1911, grant ranchers had become major architects of New Mexico's conservation effort, which was more and more a way for large landowners to privatize the hunting grounds and their game.

Statehood was granted the next year, and with it came a wildlife windfall for large landowners. New laws stipulated that any person who wanted to "establish a park or lake for the purpose of keeping or propagating and selling the game or game fish therein or to be placed therein" could apply for a license to do so. Owning such parks—Class A game reserves—meant that "all game and game fish" on the property, "with the natural increase thereof," were "the property of the licensee" to the extent "that he may lawfully retain, pursue, capture, kill, use, sell or dispose of the same in any quantity."

Unlike other states, New Mexico's game laws now allowed game park owners to own wild animals already on the land when the park was established. By acquiring land, putting a fence around it, and buying a fifteen-dollar park license, one received title to the game on that land as well.[18] Undoubtedly the legislation was designed with ranches like the Adams and Bartlett Cattle Company in mind. The essence of European game law, private ownership of game, was now a crucial plank in New Mexico's conservationist platform.

A comparison between New Mexico and Pennsylvania underscores how similar tensions and dynamics in the two areas had very different results, in part because of divergent regional histories. Crucial contrasts between the eastern state and the western territory point to the origins of a strong federal presence in game protection, which came to characterize not only New Mexico but other western states also. Both New Mexico and Pennsylvania used game laws to defend the prerogatives of landowners.[19] But restricting the prerogatives of hunting by non-landowners, as Pennsylvania did, and giving practically all power over hunting to landowners, as New Mexico did, were very different measures. If Pennsylvania could build a state commons from a base of urban sport hunters and rural landowners, why did New Mexico's wildlife commons develop a curious and seemingly contradictory dependence on both private landowners and the federal government?

The answer lies partly in demography. New Mexico's paucity of resident sport hunters created large problems for its conservationists. In Pennsylvania, as in other eastern states, urban hunters from cities like Pittsburgh and Philadelphia provided a balance to rural landowner interests; in turn-of-the-century

New Mexico, there was almost no urban base for a recreational hunting class. This was no small shortcoming. An urban recreational hunting constituency could be indispensable to creating a powerful state authority in the hunting grounds. Thanks largely to the lobbying and financial support of its sportsmen, in 1912 the state of Pennsylvania sold more hunting licenses than there were people in New Mexico, bringing $300,000 into Game Commission coffers. In New Mexico, a small population of sport hunters meant relatively few license sales and a paltry game department budget of $14,500.[20] Pennsylvania game authorities could move away from private landowners by 1912, spending proceeds from license sales to buy lands for state refuges. New Mexico, having no comparable base of urban hunters and almost no money, had to rely on private game reserves.

For all the differences between the two states, there were telling parallels as well. Just as urban hunters were the backbone of Pennsylvania conservation, the landowners who dominated New Mexico conservation were not entirely rural. Paradoxically, many of the ranching interests in New Mexico were actually extra-local urban elites—the Bartletts of Chicago and the Maxwell Land Grant owners in Amsterdam. The principal difference between Pennsylvania and New Mexico was not that one did not have an urban elite and the other did, but that, in Pennsylvania, the poles of urban and rural hunters were contained within state boundaries. In New Mexico, the geographical and cultural distance between extra-local recreationists and local subsistence and market hunters was far greater, with urban elites seeking to reorder the hunting grounds from their bases in cities that were a continent or even half a world away. In this sense, New Mexico conservation had a distinct colonial tinge. In contrast to Pennsylvania's conservationists, who served a constituency of Pennsylvania sportsmen, New Mexico conservationists appealed to national and even international constituents.

Yet, for all the collusion between state authorities and grant ranchers in New Mexico, the state was not partial to private landowners exclusively. The same structural weaknesses that created reliance on private landowners paradoxically encouraged dependence on the federal government. With little money to lavish on luxuries like game refuges, the state saw any large rural landholding as a possible wildlife refuge. The federal government owned vast stretches of rural New Mexico: between 1892 and 1912, the United States had transferred millions of acres of New Mexico public lands to the federal Bureau of Forestry and its successor, the U.S. Forest Service.[21] In 1912, the state proposed creating a game refuge in the Pecos National Forest. The state's willingness to create a refuge on public forests while endorsing private game refuges may seem contradictory in retrospect, but it expressed a coherent policy: to use large landowners—federal and private—for game conservation as a way of avoiding the expense of purchasing and managing state lands.[22] When Aldo Leopold began agitating for wildlife conservation in the national forests, he found enthusiastic allies at the state Department of Game and Fish, where his proposals echoed their own from years past.

Within the state, the visible consequence of these underlying realities was that the western state's hunting grounds came to look less democratic than the East's. Far more than Pennsylvania, New Mexico committed state power to the defense of landowner privilege. The praise New Mexico authorities heaped upon private parks and their owners suggests how much the state had come to depend on them by the early teens. In 1914, Colfax County's private parks included the William "Poke" Smith property and the Rich Brothers Ranch, but the state game warden took special interest in the "magnificent domain" of the Adams and Bartlett Cattle Company: "This property includes between four and five hundred thousand acres of the finest timber, grazing, hunting and fishing grounds in the State and is entirely enclosed by fence." Between three and four thousand of Bartlett's deer grazed the enclosure; its five lakes and two rivers ran with his fish.[23]

In 1918, state game warden Roualt noted with approval that "a great number of the larger land holders" had created their own game reserves. Roualt went on to commend the owners of the Bartlett Ranch, the Springer Ranch, the Webster Ranch, the Stern Land and Cattle Company, and the William H. Smith Ranch. "These five large cattle ranches adjoin one another and probably cover 750,000 acres or more," so that much of the county was "nothing less than an immense game preserve."[24] With large ranchers in other parts of the state now buying Class A park licenses and taking "active steps to increase the game supply . . . by the proper policing of their large holdings from poachers," private parks had become a dominant feature of New Mexico conservation.[25]

New Mexico authorities were inevitably defensive about the private parks, and they rationalized private wildlife as essential to the production of public game. State game wardens were fond of pointing out that any animals that escaped the fenced preserves became public property.[26] (The converse went unmentioned: any game that found its way onto the wealthy estates automatically became private property.) Ostensibly, private parks benefited not only state hunters but the state economy. According to official reasoning, the abundance of game flowing out of the parks (many of them ringed by two or more fences) would attract hunter tourists from across the country. Their demand for services would then redound to the benefit of local economies and local people. They would also be a boon to the state game department: if there were few resident hunters who would pay for licenses, perhaps tourists would help to pay the expenses of game wardens and restocking programs.[27] It may have been a dim vision of a distant future, but authorities in New Mexico were driven by a dream of abundance in which forests teemed not only with game, but with tourists, too. Specious as the rationalizations might be, shortage of state funds meant an absence of state power, a void that left "democratic" resource management in the hands of powerful private interests. Almost inexorably, state conservationists turned to extra-local, out-of-state interests—the grant's wealthy absentee landowners and middle-class tourist hunters.

But to some, and probably to Aldo Leopold, the close connection between private wealth and state conservation smacked of corruption, no matter how many tourists it eventually attracted. Many thought that the huge estates with their private game herds alienated what had long been a public resource, enriching the wealthy at the expense of the taxpayers. Especially galling to these critics was the state's elk restocking effort. By the turn of the century, New Mexico's elk had vanished due to heavy market hunting and competition for grazing areas from domestic stock. The territorial game warden, Thomas Gable, thought this unfortunate. He discovered (as the Bannock Indians already had) that the state of Wyoming regarded elk "as a valuable asset to the commonwealth, from a purely financial standpoint." Gable was intrigued that elk attracted tourist hunters who paid Wyoming's twenty-dollar non-resident license fee in large numbers. He decided to re-introduce elk to New Mexico, buying them with territorial money in Colorado and transporting them to suitable range in the territory. His efforts quickly came to fruition, and in May 1911, twelve elk arrived. Gable divided them into several herds, placing each on its own range.

For a public project, the effort was characterized by a strong private interest. Gable was a close associate of the Santa Fe Ring, bought the elk from an adherent of the Ring, J. B. Dawson, and placed at least one of the herds on the estate of another, George Webster's huge Urraca Ranch.[28] Webster was grateful, calling the elk restoration project "one of the best things ever done by any game warden that we have ever had."[29] Others were less complimentary. At least one critic ridiculed Gable's elk program as less a "game propagation" project than an effort to ensure that "his particular friends will have fine sport at no expense."[30] Compounding the conflict of interest was Gable's subsequent campaign to make these elk—nominally still "public goods"—into private property. The Urraca Ranch was a Class A game reserve, a private hunting estate. The 1912 statute (which Gable supported) made any game in a Class A reserve the property of the owner. The elk that Gable put on Webster's ranch became, along with their offspring, the property of George Webster. After 1912, Webster could kill or sell as many of the elk as he wished, at any time of year; he could (and did) surround his burgeoning elk herd with a fence to keep them on the ranch—and unwelcome hunters off. Gable's "elk restoration" program began by bringing elk back into the territory as public goods in 1911; it ended not by enhancing the state commons, but by enclosing a small part of it.

Leopold's campaign for public hunting grounds on the national forests was a challenge to this advancing privatization of New Mexico wildlife. Young but not foolish, he criticized the Maxwell Land Grant company more implicitly than explicitly, calling for a "free" hunting grounds. Leopold worried that game scarcity and the decline of open hunting grounds reinforced each other. At the national level, he was an ardent opponent of market-based approaches to game restoration, proposals that would allow all American landowners the privilege, recently granted in New Mexico, to own the game on their estates. The ostensible purpose would be to let owners breed game animals, slaughter them, and sell them across state lines. With the profit motive at work, the argument went, entrepreneurial game farmers would prevent extinctions. In the long run, private game would ensure the survival of public game: whenever wildlife in a park or forest declined, the area could be re-stocked with animals purchased from private refuges.

Leopold reserved his greatest contempt for these "radical game farmers." Making game private property, he argued, would turn America's conservation regimes into a copy of Europe's notoriously undemocratic game codes, where landowners owned the wildlife, hunting had become a privilege of the wealthy, and wildlife had practically vanished because the masses took no interest in protecting it. To adopt "the European system," warned Leopold, would not only be "undemocratic, unsocial, and therefore dangerous," but would threaten the survival of American game.[31]

The only way to protect wildlife was to preserve free hunting, and the only way to protect free hunting was to preserve wildlife. Leopold's strategy for doing this was to bring wildlife management to the national forests, "The Last Free Hunting Grounds of the Nation," as he called them in a 1919 article.[32] The primary authority for regulating the hunt lay with individual states. But the federal government could assist in law enforcement, and, with federal predator control programs, the national forests would become a refuge not only for game but also for hunters seeking respite from the drudgery of urban living and the enclosure of rural space. As Leopold told the Albuquerque Rotary Club in July 1917 in a lecture about wild animals, "We conceive of these wild things as an integral part of our national environment, and are striving to promote, restore, and develop them . . . as a tremendous social asset, as a source of democratic and healthful recreation to the millions of today and the tens of millions of tomorrow."[33]

Even before Leopold began his campaign, the power of New Mexico's large landowners had not gone uncontested. Despite a paucity of sources, patterns of local resistance to the private parks in northeastern New Mexico can be discerned. Working-class immigrants challenged landowner dominance in the hunting grounds: In 1909, game warden William Griffin reported that pheasants stocked by private interests in Colfax County "have not materially increased owing to the fact that the Italian miners from Dawson have killed them."[34] In 1918, a rare year in which the state game warden published the names and hometown of every person arrested for game violations, ten Italians from Cimarron were arrested for killing quail out of season.[35] Among older residents, too, resistance to the enclosure of the local wildlife commons continued for years. In a sense, the creation of the private parks was part of an ongoing process of dispossession. Since the 1870s, owners of the grant had fought with settlers who had assumed that the lands were part of the public domain and open to homesteading. When the Supreme Court confirmed that the land was in fact private and not public, hundreds of families lost their claims and faced relocation, a prospect which they often met with fury and even violence. The most bitter phase of local homesteader struggle was over by 1890.[36] Afterward, fenced game reserves helped consolidate the power of the Maxwell Land Grant Company over the local landscape. Residents of the region did not always acquiesce in the demise of older common hunting areas on the grant. In the early 1930s, the Maxwell Land Grant Company complained to the state game warden that "the Mexicans and Indians of Taos" were poaching deer in the company's Cimarron Canyon.[37] By that time, resistance to the private parks had become entrenched in a tradition of poaching, leading one observer to complain about "a certain class of killers . . . eternally at war with owners of ranches where there is game which they are not at liberty to kill as they do on areas to which they may have access."[38]

More generally, the growth of private privilege in game fueled public resentment of the grant owners. In 1918, game warden Roualt noted that "the average citizen . . . believes that the preserve owners should be forced to observe the same seasons and the bag limits as he. There has been some feeling aroused over this situation in some localities."[39] Sources from the 1930s hint at enduring conflicts between grant landowners and hunters across the state. The New Mexico Game Protective Association, a sportsmen's organization (see below), was almost invariably friendly to the private park owners. Yet even the association's literature suggests that sport hunters in northeastern New Mexico were divided over the issue of private prerogatives in game. "They realize the two sides to the large private estates being closed to public hunting. Balance in good and bad undecided."[40]

Seeking to correct the landowner bias of the game laws, Leopold exploited the growing power of urban resident sportsmen as a balance to the rural interests of ranches and wealthy out-of-state investors. In March 1916, Leopold led a convention of New Mexico's sportsmen, gathering them into the New Mexico Game Protective Association. A top priority of the 1,000-member organization was to create game refuges on the national forests, thereby giving the state a federal and public resource for game conservation and freeing conservationists from their reliance on private game parks.[41]

Leopold's tour of early 1916 through southern New Mexico was meant to build support for the organization's first meeting. Leopold's lectures that winter ignored the issue of private game refuges and focused instead on enhancing enforcement of more standard laws in the New Mexico game code, including closed seasons and license regulations. In essence, the federal government helped create a state commons in game by securing abundant hunting grounds on public forest. He urged his listeners to organize in cooperative associations as their eastern counterparts had done, to lobby their political leaders for better game laws, and to promote game protection. Only in this way could New Mexico's game be saved from imminent destruction.[42] There were sportsmen in the audience, and some members of the enthusiastic crowds who greeted him in 1916 may have shared Leopold's concern for the survival of public hunting.

But southern New Mexico's warm reception of Leopold's lectures that winter of 1916 was still remarkable. The Forest Service was far from popular in the region. For a generation, federal regulations had restricted traditional local uses of the land, such as grazing and wood-cutting. For the first few years of the forest reserves' existence, locals were legally banned from using them at all. People in southern New Mexico, as elsewhere, were often loath to accept this imposition of national authority.[43] By 1916, federal authorities had provided for various common use rights on the new federal holdings, but locals remained generally ambivalent and frequently hostile to Forest Service officials. According to lore in the Gila office of the Forest Service, one cattle rancher in these early days, when asked for his permit to drive several hundred cattle across a newly created forest reserve, touched his six-gun and softly replied, "Here is my crossing permit."[44] In all likelihood, residents of the southwestern part of the state saw the agency as a larger and more immediate threat to their traditional liberties than the landowners on the distant Maxwell Land Grant.

Yet here was a Forest Service officer whipping up enthusiasm for restriction of hunting privileges. It seemed almost too easy. As Leopold himself noted, "this thing" had been tried before. But if it worked for the young forester that winter, it was not because of shared resentment of wealthy ranchers in the northeastern part of the state so much as the way his rhetoric served the interests of one side in an ongoing, racially charged conflict over game in the state's southwestern corner.

Sources are scant, but evidence indicates that Indian hunters and Euro-American ranchers had clashed on the ranges of western Socorro County for over a decade. Beginning in the early teens, sport hunters from nearby cities joined ranchers in demanding better regulation of the hunting grounds. By the time Leopold arrived in 1916, the cause of federally sponsored game protection had broad local appeal.

If the growth of a European-style system of hunting privilege looked odd against the backdrop of the West—in myth a landscape of democratic freedoms—the solution Leopold crafted harkened to some of the darkest western and American traditions. In a curious paradox, Leopold's campaign for "the last free hunting grounds" complemented an older and ongoing effort to deny local Indians their customary hunting freedoms. Campaigning hard through the first two decades of the twentieth century, Aldo Leopold and New Mexico conservationists demonstrated both the potential of the "national commons" for securing democratic freedoms and its limits for ensuring social equity. Grasping this requires exploration of the struggle for power in southwestern New Mexico, a fierce confrontation which was coeval with the landowners' ascendance in the territory's northeastern corner. Its origins are old, but for present purposes, I begin in the fall of 1905.

The Commons and Its Promise

In 1905, the territorial legislature passed a new law offering private ranches better protection for the game herds on them. In New Castle, Pennsylvania, Serefano Diandrea was filing charges against game warden Seely Houk for assault and battery. But the head of New Mexico's Game and Fish Department, Page Otero, had his own problems. He complained to the Navajo Indian Agency in Arizona: "I have just received information that a party of nine (9) Navajos are in the vicinity of the Datil Mountains hunting." The Datil Mountains were in southwestern New Mexico, in western Socorro County, at the northern end of the Mogollon mountains. Here there were few large private landholdings; most settlers put their stock on the open range, lands belonging to the U.S. General Land Office and vast acreages nominally under the control of the Forest Service. The Navajos had been there for two weeks, "violating the game laws" and trying to sell venison. Otero requested assistance from the agency to stop the Navajos.[45]

To the game warden's chagrin, the number of Indian hunters in western Socorro County continued to grow in the following weeks. The sheer diversity of Indian hunters in the Mogollons was overwhelming: there were Navajos, Apaches, and hunters from numerous pueblos, including Isleta, San Felipe, Jemez, Santo Domingo, and probably others. Indians came from more than a hundred miles away to hunt the Mogollons, taking large numbers of deer and other game. Local settlers demanded action. On November 6, a resident of the small town of Datil complained about hunters from Isleta pueblo. At least ten Indians had arrived on horseback with seven wagons, at least four of which "they was expecting to carry them back loaded."[46] A druggist from the nearby town of Magdalena described the situation: "There are now more than a hundred Indians killing game in the Mountains, mostly the Mogollons and along the Frisco [San Francisco River] and immediate vicinity. . . . What WILL be done?"[47] The territorial game warden was outraged yet practically helpless to stem this tide of Indian hunters. By November, he was appealing to Washington on behalf of angry and fearful settlers, calling for a violent solution to the "depredations" of "these raiding devils."[48]

For all the idealistic connections Aldo Leopold drew between democratic freedoms and an abundant national wildlife commons, southern New Mexico's hunting grounds seemed imbued more with intolerance and fear than egalitarianism. Indeed, a generation earlier, a course of events like this might have ended in a war. Why did these events happen in 1905, instead of 1865, or 1875? By all accounts, Indians had been hunting the Mogollon mountains for decades; Apache and Navajo peoples had been hunting there for centuries. Why did the presence of itinerant Indian hunters become an issue in the twentieth century, long after the end of the Indian wars and long after the establishment and consolidation of reservations? Answering these questions goes a long way toward explaining why Aldo Leopold was so warmly received in southern New Mexico in 1916.

Game conservation in New Mexico frequently divided the wealthy from the rest of society. But more than that, it was set against a backdrop of complicated racial and ethnic politics. The cultural variety of the territory's local communities presented a different set of problems than what authorities confronted in Pennsylvania. The vast majority of Pennsylvanians were English-speakers. In New Mexico, English-speaking settlers, or Anglos, also dominated the top political offices and the upper echelons of the social hierarchy. But the territory's population was overwhelmingly Hispanic. Hispanos were especially dominant in the northern half of the territory, living in small villages spread across a vast region, in a combined population of more than 140,000 people.[49]

There were fewer than 16,000 Indians in the territory, and they were consigned to the bottom rungs of the social ladder. Within that relatively small Indian population, though, there was considerable cultural diversity, including nineteen pueblo tribes. In addition, there were approximately 3,000 Navajos and 1,300 Apaches (both Mescalero in the south and Jicarilla in the northwest).[50] Their customary relationships to wildlife were remarkably distinctive from those of Anglos, Hispanos, and even Pueblo peoples. Although Indians were relatively few in number, Indian patterns of land use figured prominently in rural politics, particularly in places like Socorro County.

Western Socorro County was a popular destination for Indian hunting parties in 1905 partly because the region sheltered a great deal of game, especially deer and antelope. The Plains of San Augustin, at the north end of the Mogollons, were the principal antelope habitat of southwestern New Mexico. In the late nineteenth century, white settlers marveled at the huge herds that raced across the plains.[51] In the mountain valleys of the Mogollons, mule deer and the small Sonora or fantail deer were plentiful.[52]

The abundance of game was partly a consequence of historic human uses of the landscape. Until the 1870s, the Chiricahua Apaches had been a consistent threat to settlement of the Mogollons by any other ethnic group, and periodic troubles with some Apache and Navajo bands precluded settlement by whites or Hispanos through the mid-1880s.[53] Although Apaches, Zunis, Navajos, and probably other Indians used the area for hunting, their populations were relatively small, and they frequently shifted their attention to new areas when resources in one became scarce. The mutual animosity between Apaches and other Indians probably prevented any one group from occupying the area very long. Hunting pressure was therefore restrained not only by the seasonal round of hunting, gathering, and raising crops, but by warfare. The northern Mogollons, in what became western Socorro County, were essentially a buffer zone between Apaches and others. The mountains were effectively a game refuge.[54]

Large game populations and remoteness from large settlements distinguished the Mogollons from other mountain ranges in the region. By 1900, New Mexico's northern mountains, such as the Sangre de Cristo range, had been home to Hispanic settlers for centuries, people who lived throughout the mountains in small villages. Year-round hunting of both male and female animals—sometimes for market but mostly for subsistence—was a way of life in Hispanic villages, and it helped to keep game populations down.[55] True, many Indians had also hunted year-round, but by the second half of the nineteenth century there were far more Hispanos in the Sangre de Cristos than there ever were Indians in the Mogollons. Even in the years since the Apache wars, settlement in the Mogollons in 1905 had not reached the level it did in the Sangre de Cristos by the middle of the nineteenth century.[56]

Hunters were many, but an even greater threat to northern wildlife was livestock. Sheep came north with Spanish colonists in the sixteenth century, and by the nineteenth century there were large sheep herds throughout the Spanish culture area, many of them in the north. Those numbers increased with the approach of the twentieth century. In 1880, there were almost 2.1 million sheep in New Mexico; by 1900, there were 3.5 million, and in much of the northern mountain country there were more than sixty sheep per square mile.[57]

By contrast, the Apaches did not keep livestock, and their resistance to Euro-American encroachment kept livestock herds at bay in much of western New Mexico. In the 1870s Solomon Luna, the sheep baron of central New Mexico, attempted to graze sheep in the high meadows of the Mogollons, only to retreat when Chiricahuas killed his herders and scattered his sheep in their thousands.[58] Into the twentieth century, the Mogollons were home to far fewer sheep than the northern ranges, often fewer than twenty sheep per square mile, when the northern ranges had upward of one hundred.[59]

The heavy stocking of the northern mountains pushed northern residents far afield in the search for deer. When competing with domestic stock for food, deer retreat to higher ranges. But because sheep also used the high mountain divides and because they ate the same forage plants that mule deer relied on, sheep in the northern mountains severely limited deer range. When heavily stocked ranges were combined with harsh winters, the northern mountains became relatively poor deer habitat.[60]

The Mogollons in western Socorro County were not nearly as problematic for deer. There were few sheep, and winters were not as severe, with lighter snowfall and warmer temperatures. Food was abundant in the live oaks and mountain mahogany, and, with fewer harsh winter storms, the animals could remain evenly distributed throughout the mountains for much of the winter.[61] By the early 1900s, these mountains in western Socorro County were the premier deer hunting areas of New Mexico.[62]

Many of the conditions that made western Socorro County such a superb game habitat were beginning to change by the turn of the century. Market hunters had been taking thousands of deer from some of the Mogollon valleys since at least the 1880s.[63] Livestock were beginning to make inroads in the region's game habitat, and the lower valleys were already showing signs of overgrazing. The same conditions that had suppressed deer populations in the Sangre de Cristo and other ranges were beginning to reshape the deer habitats of the Mogollons.

Still, hunting there was very good in 1905, and good hunting drew Indian hunters. At the turn of the century, all of New Mexico's Indian communities relied on hunting to fill complicated economic and cultural requirements. In general, Indians and Euro-Americans maintained radically different ideas about wild animals and what constituted proper behavior toward them. Indian hunting practices and traditions effectively created a patchwork of local commons regimes in the Mogollons, where hunters and hunted interacted in ways strikingly different from those dictated by the legislature's game laws.[64]

For Indian hunters, hunting was part of a dialogue between this world and the next: Indian hunters could retain hunting prowess only by propitiating the spirits of game animals and the powerful spirits of the predator animals from whom emanated all hunt magic.[65] Successful hunting required the observance of rituals which were often "owned" or controlled by local spiritual authorities. Among the pueblos, there were secret societies devoted entirely to keeping the hunt magic, without which no one could kill game animals. Frequently, the making of hunting fetishes and prayer sticks was the exclusive purview of hunt society members. In some places, the only people who could teach hunting prayers or sacred hunting songs were members of the hunt society, and communal hunts to distant locations like the Mogollons occurred only under their direction.[66]

As an arena for the implementation and accrual of male power, hunting was an essential part of masculine identity in local Indian communities. At Santa Ana pueblo, hunters acquired beneficent supernatural power, *ianyi*, merely by hunting. As one Indian commentator in the 1930s put it, "A hunter that tramps through the mountains looking for deer, and going through all of the hardships, must be getting *ianyi* even though he doesn't kill a deer."[67]

Although the hunt was a male activity, the entire community participated in it, insofar as proper conduct by those at home was essential for success of hunters in the field. Animal spirits could see impropriety in the pueblo, and offending them guaranteed that they would not offer themselves to pueblo hunters. Children at home strove not to be "mean" while relatives were hunting lest the deer be "mean" to the hunters.[68] While their husbands were away, wives of pueblo hunters avoided even the suggestion of infidelity (such as talking to single men), because infidelity at home brought failure or misfortune to hunters in the field.[69] Men dominated the act of killing, but securing game and propitiating the spirits was a community endeavor.[70]

The bounty of the hunt reinforced local community ties. All game killed on pueblo communal hunts belonged to the pueblo except that killed on the last day, which belonged to the hunter.[71] Game was an essential bond among non-pueblo communities also. Among Navajos, meat was divided among the members of the hunting party.[72] Apaches required their hunters to be generous, sharing the meat and even giving it away entirely to the first person who asked for it.[73]

The complicated social and religious activities that supported hunting did not preclude it from becoming a moneymaking enterprise. Indeed, economic factors reinforced the cultural importance of the hunt for Indians. The early 1900s were an economic nadir for many Indian communities.[74] When there was a market for meat or hides, however illegal, hunting was a way of earning cash. When there was no market to be found, hunting helped people to live without money. At the turn of the century and beyond, hunting provided food, skins for leggings, moccasins, and shirts, as well as feathers for ceremonial purposes.[75] After decades of seeing their land stripped away and their sovereignty eroded by the Euro-Americans who dominated the territory, Indian people found that hunting provided both symbol and substance of independence.

Despite their similarities, Indian beliefs and hunting traditions often differed from one community to the next. In general each Indian community regulated its own interaction with game animals and the land, relying on prayers, rituals, traditions, and technologies which—however similar they may have been to other pueblo practices—were within the purview of local agents and authorities. Varying hunting practices created different local hunting patterns, distinctive local commons regimes.

The territories of these various local commons regimes often overlapped. Historically, it was not unusual for Indian communities to make conflicting claims to the same hunting ground some distance from home.[76] In most cases, these overlapping commons areas were the destination of communal hunts rather than individual hunters, since distance and the possibility of encountering enemies on such a journey militated against embarking alone. After the turn of the century, the Mogollons were an attractive destination for many communal hunters.[77] Many Indian hunters apprehended there in the early twentieth century were in groups numbering between ten and thirty. Communal hunts were conceivable at any time, but autumn, when deer and antelope were fatter and their hides in better condition, was an ideal time for a foray into the Mogollons. There a large amount of game could be had in a week or two. With so many Indians hunting the area, it is impossible to locate a discrete "local commons" there after the 1880s. In fact, it was less a bounded region with an exclusive group of ethnically or culturally defined users than a hunting locale with relatively fluid borders. At times, it probably approached the status of an open hunting grounds, perhaps subject to more heavy use than a single local commons regime would have allowed. That said, Indians who hunted there often had long traditions of mutual hostility which probably militated against simultaneous use of any area. Moreover, they were loath to give up what they had come to understand as customary prerogatives there. Whether or not the region was a unified local commons, the hunters who used it thought of it as a crucial part of local life, and even a local right. So in 1905 the Mogollons were a locus of overlapping local commons regimes, a landscape of encounters between occasionally similar, sometimes related, but always distinctive local cultures.

The coming of the conservationists meant new challenges to Indian hunters in the Mogollons and elsewhere in the territory. In contrast to Indian traditions, which vested authority in local spiritual powers, conservation placed unprecedented authority in extra-local secular powers. Around the turn of the century the legislature sought ways of reshaping the hunt in southwestern New Mexico at the same time that it aggrandized the owners of the Maxwell Land Grant. Beginning in 1895, the territorial legislature imposed a nine-month closed season on deer, elk, and antelope and restricted the sale of game meat.[78] Other game laws followed shortly thereafter. The 1901 law outlawing the hunting of big game like deer on public lands had obvious ramifications for Indian hunters. To enforce the new game laws the 1903 legislature created a Territorial Department of Game and Fish, a one-man operation administered by the territorial game and fish warden.[79] The legislature soon repealed the deer-hunting ban of 1901, but in 1905 it outlawed the killing of does and stipulated that "deer with horns may be killed with a gun only during the period commencing September 15, and ending October 31, of each year." The same law limited each hunter to one deer per year.[80]

Game laws like these indicated differences in how Indians and conservationists thought about wildlife abundance. For Indian hunters, game would appear if hunters behaved properly and propitiated the spirits; to conservationists, game would become abundant only if state or territorial authorities controlled access to the hunting grounds and limited the take.[81]

But in the early 1900s, the legislature's legal challenge to the local commons was perhaps less immediate than other, more material changes in the hunting grounds themselves. Once the Apaches were forced into reservations in New Mexico and Arizona, new settlers quickly moved to fill their old homeland. A rising tide of Euro-American settlers engulfed the foothills of the Mogollons as the region's mining matured in the late nineteenth century, and by 1900 more and more rural settlers were staking out claims for ranches and farms.

The immigrants brought with them new ways of interacting with the land: Hispano stockmen immigrated from the north and east, while Anglo cattlemen pushed up from the southeast, especially Texas. By the 1890s, Socorro County and indeed all of western New Mexico was undergoing a rapid increase in population and stocking rates.[82] The population of Socorro County grew by 27 percent between 1890 and 1900. Grant County, the civil division that bordered Socorro County on the south and from which many Anglo hunters ventured into western Socorro County, had an even bigger increase, at 33.4 percent. Significantly, most of the growth was in smaller, outlying precincts, rather than in the larger settlements of Silver City, Socorro, Magdalena, or Kelly.[83]

This marked increase in Euro-American peoples had several important effects on the patchwork of commons regimes throughout the Mogollons. First, the settlers brought with them new market ties between the Mogollons and distant urban settlements in their cattle, sheep, and other stock. Even smaller homesteads had a few sheep, goats, cattle, or horses. The establishment of a railhead at Magdalena opened an outlet to national markets, further encouraging livestock production and sale. By the turn of the century, there were 150,000 cattle and 250,000 sheep grazing the open ranges of Socorro County.[84] In the minds of settlers who turned their animals loose to roam the fenceless hills and plains of the county, the region was more useful as open range, a settler grazing commons, than a hunting grounds.

Just as the livestock market was a force in shaping the interaction between new settlers and the hunting grounds, so it reshaped the land itself. Cattle probably had little effect on deer habitat, because deer could retreat to the higher mountain slopes, especially in the summer. But at lower elevations livestock had a profound impact. The expansion of sheep onto the Plains of San Augustin around 1900 wrought such changes in grassland communities that a range inspector in 1907 reported a marked decline in blue grama grass, the preferred grazing species and the predominant grass of the Plains. In some places blue grama had been almost completely replaced by sheepweed, an unpalatable grass that succeeds where other species have been grazed beyond their capacity to regenerate.[85]

Changes in the physical composition of the grasslands were most visible in the decline of antelope. The Plains of San Augustin were the principal antelope habitat of southwestern New Mexico.[86] Antelope do not compete well with livestock on heavily stocked ranges; between the rising numbers of livestock and the increase in resident hunters in the valleys at the foot of the Mogollons, antelope herds began to shrink much as they had in other parts of the territory after settlement. In October 1905, a biologist from the Bureau of Biological Survey described antelope of the San Augustin Plains as "formerly very abundant" but now "sadly diminished in numbers."[87]

To local settlers, the decline of antelope was a portent of ominous shifts. The huge antelope herds of the Plains of San Augustin were legendary among locals in the region by the end of the 1880s, when they were already in decline. Some blamed their demise on severe winter snows in the latter part of that decade.[88] But others blamed Indians. In 1908, biologist Vernon Bailey recorded a conversation with Hank Hotchkiss, a local rancher. Hotchkiss complained of Navajo Indian hunters on the Plains that year, one of whom tried "to stalk a bunch of antelope by holding before him a canvas stretched on a hoop and painted like an antelope." Some nearby cowboys fired at the antelope, scaring them away from the Navajo hunter, who went after them. According to Bailey, "The settlers in the region complained that these hunting parties were largely responsible for the rapid decrease of game."[89]

For all the divisions between Indians and local settlers, rifts within settler society itself shaped the confrontation with Indian hunters. The surge of Euro-American population in Socorro County was not a unified incursion, but the simultaneous expansion of two distinctive and often hostile cultures. The northern and eastern regions of Socorro County were over 90 percent Hispanic in 1900, as was Valencia County to the north. The population of the rest of Socorro County was more Anglo, and the southwestern quarter of the county was strongly Anglo.[90] For Anglo cattlemen and Hispano sheep raisers, the early 1900s was a period of intense and sometimes violent competition for range land in much of the territory.[91]

In this contest for land between rival Euro-American ethnicities, Indians were shunted aside. Both Anglos and Hispanos grabbed land from Indians, whose claims were ignored in the tense disputes between new settlers.[92] And New Mexico's nascent game laws became a valuable tool of settlers seeking ways to push Indians off their customary hunting grounds.

Game laws helped create a new social order on the land by demarcating a new wildlife commons, one that fit the socioeconomic needs of settler society better than the older commons of the Indian communal hunts. Whereas an Indian community viewed land and wildlife as one local commons—a commons as place—Euro-American settlers set about individuating land and abstracting wildlife into a pool of resources owned in common. Nominally, wildlife was protected by the state; in day-to-day reality, it usually remained the de facto property of settler communities.

This fact was apparent in the way that Socorro County's new residents campaigned for law enforcement against Indians but refused to adhere to game laws themselves. Anglo and Hispano immigrants to the area were miners, ranchers, and farmers—people who were more comfortable with subsistence and market hunting than with notions of "sportsmanship" or conservation. For Hispanos who immigrated from the northern mountains, hunting year-round was a long and venerable tradition. Game-law violations were similarly pervasive among Anglos. According to one Anglo resident of the Mogollon mining towns in 1904, "If there were any game laws then no one observed them; first one and then another would bring me a piece of venison in all seasons of the year except mid-summer when it would not keep so well."[93]

In this respect, there were marked class differences between Socorro County's settlers and the upper-class champions of New Mexico wildlife law. As in other regions, conservation in New Mexico had been the cause of recreational nature enthusiasts, people like Page Otero, the first territorial game and fish warden. An active angler and hunter, Otero was also a scion of one of the wealthiest and most powerful of New Mexico's families (his brother was the territorial governor) and a graduate of Notre Dame. In office, he corresponded with prominent American sportsmen, and he attempted to organize New Mexico's recreational hunters in much the same way Leopold did more than ten years later.[94] Among the broad population of subsistence and market hunters in southwestern New Mexico were probably a few conservationists, the "Silver City Sportsmen" whom Otero credited in his 1903 report for "their great interest in protecting and preserving the game and fish" in southwestern New Mexico.[95] But the presence of sportsmen merely ensured that even within Socorro County's settler society there were class divisions with conflicting perceptions of what constituted a proper relationship to wildlife.

For these disparate class and ethnic interests, Indians provided a unifying target. The mass of rural settlers and the small number of elite conservationists could set aside their differences over wildlife, coming together behind the cause of game protection because of the supposed threat that Indian hunters posed to the settlers' hunting and livestock. The years immediately preceding 1905 were apparently fraught with tension between Indian hunters and white settlers. The game law of 1901 specifically applied "to all Indians on or off the reservations, or coming into this territory from adjoining states or territories."[96] A similar provision was included in the 1903 game law.[97] In his reports of 1903 and 1904, the game warden excoriated Indian hunters for either-sex deer hunting, market hunting, and wastefulness. "The slaughter of deer by the Apache, Navajo and Pueblo Indian tribes who make their raids from their reservations each fall (and at other times) has done more than all other agencies combined to reduce the number of deer and antelope herds."[98]

Settlers joined Page Otero in blaming Indians for the decline of game, but they were at least as concerned for the safety of their livestock. In early 1905 Otero began an effort to clear the range of Indian livestock as well as Indian hunters. The game warden sponsored a joint resolution of the territorial legislature, demanding that the U. S. Department of the Interior take action against Navajo herders and poachers. Not only were Navajos "constantly violating the game laws" and "trespassing upon the public domain and the property of American citizens" over almost all of western New Mexico, including Socorro County. They were also frequently "going off their reservations and grazing their stock on the lands of the citizens, driving the citizens stock from off their own lands and in some instances stealing and destroying the stock of citizens."[99]

In fall 1905, the situation was no better, particularly in Socorro County. "Every day I am receiving letters from the settlers there complaining of the Indian depredations," Otero wrote in one of a series of letters to the Navajo Indian agent. "In many instances when game was scarce they have killed cattle and sheep."[100] Later that season, Otero wrote to the Department of the Interior, complaining again about Navajos whose range conflicts with white settlers had continued for more than a decade.[101]

Other settler concerns probably aggravated their worries about Indians. These were nervous years in ranch country. The livestock market, in reshaping the land and people's relations with it, was creating new forces of change. Heavy grazing decimated not only the Plains of San Augustin, but much of New Mexico. In every region of the territory, ranchers noted the disappearance of the preferred livestock grasses, such as perennials, and their replacement by annual six-weeks grasses and weedy forbs and shrubs, all less nutritious or palatable for stock.[102] In interviews with a range examiner, stockmen described the erosion of soil that resulted from heavy rains in 1905 and a loss of plant cover. "They all say that years ago the ground was level enough to drive over with a wagon where it is now almost impassable for a horseman. They also claim that certain areas were once cut over for hay, while there is now little or nothing on them, much less a hay crop." The ground, barren of grass and trampled by millions of hooves, no longer absorbed water as readily as it once had. Small streams, once reliable, now cascaded briefly after the rains then went dry. Many pools and springs diminished or disappeared entirely.[103] Combined with a livestock market that took a downward turn in the early 1900s, the eroded land made ranching prospects seem dim.[104] Sheep population declined with the exhaustion of the range, and cattle numbers remained stagnant.[105]

Confronted by these challenges of fickle markets and changing land, many ranchers who depended on the range for their livelihood sought ways to control, manage, and rationalize it. Across New Mexico, stock raiser relationships with the open-range grazing commons began to change. Large sheep ranchers turned to sedentary grazing, carefully controlling the numbers of sheep and strategically depasturing them to ensure continuous availability of grass. But the most noticeable consequence of this push to regulate the range was fencing. As the number of livestock on New Mexico's ranges stagnated and began to decline, between 1900 and 1914, ranchers across New Mexico began erecting fences in earnest.[106] In the Datil area of western Socorro County, according to one commentator, "Everybody fenced in as much territory as he could afford barbed wire to stretch around."[107] Fences allowed ranchers to control stocking rates, keep neighbors' stock from competing with their own, and to rotate stock from one area to another, providing the grass crucial rest. As one contemporary observer put it, control of the land "changes a 'range' into a set of 'pastures,' and the presence of the fence eliminates a very large proportion of the uncertainty of the business."[108]

Although in many cases ranchers fenced land that was not theirs (a tendency that sometimes led to bitter fights), fences created de facto private parcels from the old common grazing area. And, as in Pennsylvania, the demise of the commons as place was accompanied and reinforced by the creation of an abstracted state commons in wildlife, in which game became a resource pool separate from land, and game laws stood as figurative fences between wild animals and lower-class hunters. The fact that antelope and cows grazed on the same land meant an overlapping of wildlife and livestock pasture. Ordering one could help order the other. And in enforcing game laws against Indians, ranchers sought to create a new and more "ordered" grazing regime. If fences were physical boundaries on the land, expressing rancher control over property, game laws were legal boundaries, reinforcing rancher control over anonymous men with guns.

Rancher concerns about Indian hunters paralleled the worries of Pennsylvania farmers about "marauding gunners." But the Indian wars had spawned a gruesome mythology, and to New Mexico's rural settlers, Indian hunting parties were more frightening than any roving band of armed city slickers or even Italians in the Pennsylvania countryside. The life insurance company that held a policy for Page Otero threatened to cancel his coverage if he rode into the field in pursuit of Indian poachers, and as late as 1913 stories of encounters between sportsmen and gun-toting Indian hunters were front-page news in city papers.[109] If Indians with guns frightened insurance companies and urban sportsmen, they must have been doubly worrisome for ranchers, miles from each other and days from larger settlements. One of game protection's greatest appeals for rural non-Indians was its aim of "regulating" the hunt, ridding ranchers of the worrisome sight of armed Indian bands. Page Otero spoke to these pervasive concerns in his letters to rural New Mexicans, many of which read substantially like the one addressed to a Sandoval County resident in 1905: "Everybody in that country should be anxious to protect the game and fish and see that Indians especially are kept off the range."[110]

Of course, few if any settlers wanted or expected impartial enforcement of game laws. Through alliances with extra-local authorities like Page Otero, "new locals" (the settlers) often defended a new local hunting commons against older locals (the Indians). Game laws, selectively enforced, served as an ethnic perimeter around the new settler hunting commons, in which poaching by settlers was tolerated but hunting by Indians was not.

To be sure, there were ethnic divisions within this new settler order, and the alignment of settlers on the issue of game protection fell out along ethnic boundaries. Of those who made written requests for the game warden's help in 1905, one characteristic is striking: all seem to have been Anglo. Hispanos may have filed complaints about Indian hunters with Page Otero by word of mouth. But if they did, no record of those complaints survives. This in itself is remarkable, because Hispano stock raisers were a major component of Socorro County society; over the northern half of the county, Anglos comprised no more than 50 percent of the population.[111] Yet, to judge by the evidence, calling in the territorial game warden was overwhelmingly an Anglo recourse, not a Hispano one.

The predominance of local Anglos among supporters of the territorial game warden probably paralleled social reliance on the other boundary-marker on the commons, the fence. Historically, fencing the range was an Anglo practice.[112] We cannot know how many Hispanos erected fences in Socorro County at this time, but the fact that the wave of Hispano immigration from the northeast was comprised of expansionary sheep herders, at a time when sheep herders were decidedly nomadic and hostile to fencing, suggests that fencing was not a priority in most Hispano communities. The bounding of the land in Socorro County reflected not just Euro-American power, but Anglo power; Anglos turned to central authorities and fencing to apportion local resources in accordance with their own ends.[113] In the construction of a new relationship between central authority and local commons, Anglo locals formed the crucial bridge between territorial authority and local settlements.

Given the pronounced ethnic conflicts over wildlife conservation elsewhere, we might expect that Anglos persecuted Hispano poachers as well as Indians. Game conservation in New Mexico remained an overwhelmingly Anglo movement for decades to come.[114] Many conservationists probably would have liked to use game laws against Hispanos as well as Indians. But in a county with such a large proportion of Hispanic residents, at a time when the territorial game warden and the governor were Hispanos, and when Hispanos were the majority of registered voters in the territory, arresting large numbers of Hispanos hardly ensured the success of game laws in future legislatures. Selective use of territorial power demarcated and defended a new commons for the nearly exclusive use of immigrants, so long as those immigrants were politically powerful and the people excluded from the commons were politically weak. Outnumbered, impoverished, and disfranchised, Indians were easier to exclude than the majority Hispano citizens.

Indians did not soon retreat before the conservationist onslaught. Their attachment to the Mogollons as a hunting area could not be legislated away, and, consequently, Anglo perceptions of rural order were impossible to implement without considerable support from territorial and federal authorities. By themselves, early territorial measures failed to restrict Indian hunters, as the course of events in 1905 demonstrated.

In the fall of that year, having received a number of complaints about Indians hunting in western Socorro County, game warden Otero began seeking a means to bring the law to bear on them. Under territorial game law, Otero could appoint local deputies who had powers of enforcement. But, as he complained, the Indians "pay no attention to my deputies." Indian resistance forced the Department of Game and Fish to rely on two other government agencies: the newly created New Mexico Mounted Police and the U.S. Forest Service.

Whereas sheriffs and constables had power only within their own counties, the eleven officers of the New Mexico Mounted Police had authority to enforce territorial laws anywhere in the territory.[115] Modeled on the Texas Rangers, the primary purpose of the mounted police was control of the range through apprehension of rustlers.[116] Otero sent the police to western Socorro County to forestall Indian hunters (who were widely believed to be rustlers as well as poachers), warning the Navajo Indian agent of "serious trouble" if the poaching continued.[117] At least in Otero's mind, the campaign against Indian hunters quickly began to resemble the buildup to an Indian war. "I am heartily glad that some action will be taken against the depredations of these raiding devils," wrote the game warden to the police captain.[118] But the situation continued to deteriorate. By mid-November, Otero wrote that he had received twenty letters in the previous ten days "from settlers in Socorro [County], complaining of these Indians."[119]

For all the game warden's enthusiasm, the Indians' success at evading capture revealed the weakness of territorial authority. On November 16, Otero received word that Lieutenant Cipriano Baca had apprehended a party of San Felipe hunters with four doe hides. He tried to arrest them, but the Indians refused to cooperate. According to later reports, Lieutenant Baca "said he would have to kill some of them" if he continued with the arrest. To avoid violence, Baca and the other officer released the hunters, planning to make arrests on the Indians' return to the pueblo.[120]

It was already clear that the situation in Socorro County was beyond the control of the mounted police. According to the police captain, John Fullerton, there were "in the neighborhood of one hundred and fifty Indians," from pueblos as far afield as Jemez and Isleta, hunting and on their way home from hunting in the Mogollons. Complaining that he had "so few men it would be almost impossible to arrest these Indians," Fullerton recommended that Otero take to the field himself and that he deputize ranchers to assist the police.[121]

The weakness of these authorities, who were administering a vast and thinly populated region on a shoestring, was partly ameliorated by the presence of federal officers from the U.S. Forest Service. Indeed, the evolution of New Mexico's "state commons" as a highly federalized regime owes its origins in no small degree to the eagerness with which foresters took up the challenge of expelling Indians from the hunting grounds, a proclivity that helps explain the enthusiastic response Aldo Leopold met in southern New Mexico a decade later. Rising federal authority over game was paradoxically a function of weak federal power over other forest resources; settler resistance to the Forest Service was instrumental. This dynamic is perhaps most clear in the story of the Gila National Forest, which includes much of the mountainous forest land in southwest New Mexico.

Immediately after its creation in 1899, the Gila Forest Reserve was off-limits to all users. Legally, it was a category of federal property that was not a commons at all. Of course, there was stalwart local opposition to this rigid policy. Locals continued to graze cattle, cut wood, and hunt much as they had before, in such numbers that genuine federal management of resources proved difficult where it was not impossible.[122] This was especially true where game was concerned. After rangers were empowered to enforce state game laws, their biggest obstacle was the pervasive reluctance of local officials to restrain poachers. Juries usually failed to convict Anglos, even those who violated game laws openly.[123]

In response to this widespread resistance, forestry officials began to search for a more flexible and workable management policy, one that allowed locals a measure of customary use rights. In this way, they hoped to encourage cooperation between settlers and officials and thereby strengthen national authority in the reserves. By 1905, authorities had liberalized the stocking policy, so that ranchers who obtained the necessary permits could depasture their animals on the forests under the watchful eye of federal officials. Settlers could also cut timber for domestic use. In 1905, the Forest Service began charging nominal fees for permits, and soon thereafter began the process of allotting pasture, designating specific sites where each permittee could put his livestock in the forest.[124]

By 1905, residents retained certain bundles of use rights in the national forests (federal policy in fact favored residents over non-residents in allotting permits), a policy devised by federal authorities seeking to manage the use of the forests to ensure their viability for the public. Although it was a place with legal boundaries, authorities also understood the Gila National Forest as part of a pool of resources or public goods like wood, water, grass, and game. In this sense, forest rangers and their superiors saw their job as the maintenance of a national commons in forest products.

Whether this bureaucratically administered regime could ever approach a true commons—a community property system owned and managed by its users—was open to question. But in any case, because the national forest included much of the region's wildlife habitat, conservationists' nascent state commons in game was tied to this national commons in forest goods.

Legal proscriptions reflected this overlap: by 1905, federal authorities' major tool for regulating the hunt was state game law. And, in fact, federal forest rangers were among the most ardent enforcers of state laws targeting Indian hunters, a penchant that commingled state commons and federal commons by more than geography. Paradoxically, this development occurred in large part because federal authorities were anxious to avert or ameliorate settler resentment over restrictions on timber cutting and grazing in the national forest. When local Euro-Americans insisted that they arrest Indians, forest rangers did so, in part because settler courts were anxious to convict Indians of poaching, and in part because doing so provided rangers with a bridge to local settlers. Even these divided camps—federal authorities and local ranchers— could join forces when it came to arresting Indian hunters.

Looked at from another perspective, the story reveals the shifting meanings of "local" and "national" in the wildlife commons. New locals, the immigrants, were demanding federal, extra-local action against older locals, the Indians. Ranchers complained to ranger Fred Winn on the Gila and forest supervisor R. C. McClure in Silver City about illegal hunting by Indians. "It seems to me that it is an injustice to any of us citizens of the county that the Indians be allowed in here and to kill as many deer as they are pleased to kill," wrote sheep raiser James Patterson from his headquarters on the Plains of San Augustin. "I would like to see some of these Indians taken up and put off the timber reserve and made [to] stay away."[125]

According to the recollections of Benjamin Kemp, a settler on the headwaters of the Gila River, in the early 1900s numerous Indian hunting parties camped in the area. "No one knows how many deer they killed and crippled." But in 1906, "the Forest Service began to take a hand and it was not long until things began to change for the better."[126] The struggle in southwestern New Mexico demonstrated clearly the transitory nature of local identity, as people from elsewhere, extra-locals, became "local" through the process of immigration, and then demanded national assistance to reorder or extinguish the older local commons, the Indian hunting grounds.

Indeed, territorial game authorities looked for ways to increase federal involvement further, as the assistance of forest rangers proved inadequate. Soon Page Otero was urging settlers to "take care of these matters" themselves in hope that a violent incident would spur federal action from the Bureau of Indian Affairs.[127] In late November 1905, Otero regretted that Lieutenant Baca had missed his chance to shoot it out with the San Felipes. "I think that if the settlers would kill off a few of them it might be the means of attracting a little attention from the Department at Washington and bring about the desired result. I heartily wish this would be done."[128]

The culmination of Otero's efforts was an appeal to the Department of the Interior in which he cited "reliable businessmen" (presumably ranchers) who informed him there were "twelve hundred or more" Navajo Indians living off the reservation and seizing the stock of citizens. The Pueblo Indians "are equally as troublesome as the Apaches and Navajos."[129] The game warden did not have the power or resources to resolve the growing confrontation, and he warned federal authorities that "settlers and stock men are organizing, and if something is not done soon to put a stop to this raiding of the different Indian tribes, there is likely to be serious trouble."[130]

With at least one local newspaper carrying headlines about "lawless" Indians, Otero waited hopefully for a federal response.[131] But federal authorities refused to intervene, directing that Indian poachers be treated like any other lawbreakers. Otero was incensed. In an ominous letter to the territory's congressional delegate, he wrote that he had advised settlers "that a few severe examples made of these Indians would probably then attract the attention of the authorities in Washington. . . . This would probably have more effect than anything I can do with my limited power."[132]

Thereafter, he advised numerous correspondents to shed Indian blood. "These violators must be taken one way or another," he advised one resident in northern New Mexico.[133] "If they resist you," he wrote to another in pursuit of hunters from Jemez Pueblo, "you know what to do."[134] The day after Christmas he wrote to another, "Any Indians you can find, arrest them, and if they resist, kill one or two, and then we will probably hear from their Patrons."[135]

Surprisingly, no one was killed. The communal hunts were probably over by the end of December, by which time winter had come to the Mogollons. Ranchers had other things on their minds, and there were probably few Indians to be found. Still, Indian hunters continued to be the bugbear of New Mexico's conservationists for the next fifteen years. Annual reports of the game department often discussed the "Indian problem" in game conservation, and most of the complaints seem to have been directed at Indians hunting in the Mogollons.[136]

There were new complaints in 1906, and in 1907 forest ranger Thomas Verner arrested eleven Navajos on their way home from a hunting expedition in the Black Range "with about eighty deer skins, a quantity of venison, and other evidences of their slaughter of game."[137] In 1910 territorial game warden Thomas P. Gable arrested five groups of Indian hunters, including a party of twenty-seven Lagunas. As in earlier arrests, the Lagunas were arrested in the fall, in the Datil Mountains. They had more than a hundred carcasses, deer of both sexes and all ages. And, as in earlier years, the arresting agents—a rancher and a deputy game warden—came from the ranks of local settlers and territorial officers. Federal officers also made an appearance that fall. "Valuable assistance was also rendered this department by Supervisor W. H. Goddard and his forest guards by the arrest and conviction of violators of the game laws in the Datil Forest district."[138]

Indians resisted the game laws at great cost. The San Felipes who refused to cooperate with the mounted police in 1905 were finally arrested and brought to trial in Socorro County in spring 1906, each of them held on $250 bail.[139] When Judge Green faced the eleven Navajos brought before him for poaching deer in 1907, he sentenced them to thirty days in jail.[140] The Laguna party arrested in the fall of 1910 paid $625 in fines. The penalty was paid by the pueblo; according to the territorial game warden, the hunters were to be "held under a system of peonage until they pay back this money to the pueblo."[141] The local press covered the trial of the Lagunas, conveying a sense of their hardship: "The Indians were allowed to retain the deer hides, meat and heads which they obtained on the hunting trip, on account of their impoverished condition, and to enable them to reach their homes. . . . The penalty assessed against these violators of the game law was made as light as possible owing to the condition of the Indians."[142]

Others were less sympathetic to Indians, especially to a party of Apaches arrested the same year. In November 1910, a U.S. forest ranger arrested a party of ten Apaches for hunting without licenses in the Mogollons. Tried in the town of Reserve, they were fined $500 and costs. "Being without that amount of the coin of the realm in their possession, the Indians sold their ponies and equipment to pay the fine and then promptly hiked back to the bosom of their campfires in Arizona," wrote a jubilant reporter in the *Socorro Chieftain*.[143] Two sportsmen who saw the Apaches gave a report to the *Albuquerque Journal.* Whatever the tone of their dispatch, by the time it appeared in the Socorro newspaper it was practically hysterical. "It is a regular massacre," the newspaper quoted the sportsmen as saying. "The Indians swept through the mountains like a line of elephant beaters, a regular army of them, and what they did not kill they drove out of the country. The bunch that were rounded up at Reserve weren't one tenth of the Indians in the hills." The paper reported that local residents had seen "hams and quarters of the slain deer" in stacks breast high. "A few raids like that will effectually wipe out every bit of game in the country. It is an outrage."[144]

The fact that this last report came from two urban recreational hunters reflected another change in the hunting grounds of southwestern New Mexico, a change that had vital consequences for the fate of Aldo Leopold's tour in 1916. By the early teens, middle- and upper-class recreational hunters were emerging as a political force in the southwestern part of the state (see chapter 4). They were among Leopold's most ardent supporters, and although their assumptions about proper interactions with game differed dramatically from those of rural settlers, they shared settler concerns about Indians. In much the same way that rural landowners and urban sportsmen combined against the Italian immigrants and "mountaineers" of Pennsylvania, ranchers and sportsmen joined together against Pueblos, Navajos, and Apaches. In November 1914, a posse of recreational hunters set out from Silver City, on the trail of Navajo hunters in the Mogollons. In short order, they brought a party of fifteen men to trial for illegal possession of deer.[145]

Among the problems foremost in the minds of ranchers and recreational hunters, Indians were soon to fade. Tourist hunters in western Socorro County created almost as much trouble as Indian hunters, and possibly more. Socorro County became a popular recreational destination, not least because, at last, the state game department was beginning to realize its old dream of a tourist hunting ground on federal land. The state game warden could now describe Socorro County as "one of the best game counties" in New Mexico, with "a large area of the National Forest, upon which the game takes refuge, to the great benefit of the hunter." Transportation networks that took livestock out of the range to urban markets could also bring tourists from urban markets to the hunting grounds. "The hunting and fishing grounds of the county are usually reached by train to Socorro . . . and on a branch to Magdalena, and from there good roads extend to all important points in the county."[146]

By this time, Indian hunters had developed innovative ways of resisting the game laws and persisting in their assertions of a viable local commons. Sometimes they carried letters of introduction from friendly ranchers.[147] In other ways they made more direct challenges to central authorities. Frequently they exceeded the bag limit and violated the buck law. Game warden Gable reported in 1910 that on two occasions, "Indians attempted to get the hides and heads of their booty into the pueblo by sending special couriers away from the main body."[148] Buying licenses was a way of avoiding on-the-spot arrest, and it became such a standard recourse that the state game warden complained in 1916, "The majority of them now secure the necessary license for their hunting, but without constant watching they will hunt out of season and take more than is permitted of any kind of game they can kill."[149] That year, a federal wolf hunter encountered a party of five Navajos on the Plains of San Augustin; they were leading four packhorses loaded down with deer and antelope hides.[150]

Indians remained a concern of local ranchers, sportsmen, the New Mexico Game and Fish Department, and the U.S. Forest Service for a long time to come. That concern helped to unite ranchers and sportsmen, albeit on shaky ground with plenty of potential for future conflict. But in December 1915, the regulation of the land for livestock markets and the control of hunting for the tourist market were imperative in driving metropolitan sportsmen and hinterland stockmen in the same direction. Together, they were a ready-made audience for a charming advocate of game-law enforcement on the national forests. In January 1916, Aldo Leopold stepped off the train for his first speaking engagement, in Silver City.[151]

This much was certain: the forests that blanketed the Mogollons, and the rolling foothills at their feet, were a hunting grounds. Whose they were was another question. In the claims of Anglos and Indians to this land, one can see the contested visions of the hunting grounds and the questions they posed for the new, heavily federalized hunting order that emerged on it. To Aldo Leopold the mountains represented an arena for hunters who could afford licenses and the journey there. With an abundant wildlife population, the national forests could stem the advance of private privilege in game. In this way, national forests were a "national environment," a repository of democratic hunting liberties.

Another kind of hunting freedom was enshrined in settler hunting custom, for the layering of state and national authority over this landscape never completely extinguished local claims to it. In the early 1900s, local Anglos poached game while simultaneously calling for better game-law enforcement against Indians. This local commons bounded by state law remained viable for a long time.

A glance at hunting regulation in a local community several decades later makes the point. In 1949, anthropologist Evan Z. Vogt described the complex social code that local Anglos had developed to regulate the hunt in the town of Fence Lake, a short distance north of the Socorro County line. According to Vogt, Fence Lake residents had little use for game laws, which they believed to be tools for protecting deer for "city folks." Locals considered subsistence hunting of deer, even out of the legal season, "a legitimate, economic pursuit."

Virtually all men, including the state-appointed game warden, hunted deer year-round.[152]

In this local hunting order, the real crime was not poaching but reporting a poacher to outside authorities. Whenever one local reported another for poaching, " 'all hell breaks loose' within the community and the person who 'turns somebody in' is punished by techniques ranging from gossip, through social ostracism, to 'beating them up,' in the more extreme cases."[153]

Hostility to outside authority did not mean that Fence Lake rejected all regulation of the hunting grounds. On the contrary, to preserve a share of local abundance for all local hunters, residents enforced a code of behavior. In Fence Lake, killing deer indiscriminately "means one is a 'game hog' whose actions will also be controlled by informal but effective methods," not unlike those used against people who reported local poachers.[154] This was a local commons, in which local people managed access to wildlife within their own community, despite the presence of state and national authorities. Presumably, Fence Lake residents kept outside interference to a minimum through enforcing state game law against extra-local hunters. Insofar as it expressed local self-determination, this local commons was a manifestation of local power, and local liberty.

Locals and tourists alike found access to these hunting grounds much cheaper than the entry fees for private clubs; in that sense, Leopold's dream had become a reality. But there were great costs inherent to the creation of this "free" hunting grounds. For Indian hunters, the maintenance of the old local commons became increasingly difficult with the rise of livestock and tourist markets and the ascendance of the game law. As the Mogollons were connected to the national economy in new ways, and as conservationists set about preserving their deer as goods for the public, many Anglos and Hispanos celebrated. It seems unlikely that Indians would have joined them. We cannot know what they thought about these developments, but for the Pueblos, Navajos, and Apaches of the region, this new hunting order could not have been connected to any notion of freedom.

4
Tourism and the Failing Forest

Sixty years after Leopold's campaign began in 1916, New Mexico's sporting circles were in a less triumphant mood. "Your vile, ungentlemanly, libelous letter of January 3 is received," wrote Elliot Barker, for twenty years the head of the Department of Game and Fish and now the elder statesman of New Mexico game conservation. "Your name calling and vilifications don't worry me a bit, I just consider the source and forget it."[1] His correspondent, a supporter of an organization called Sportsmen Concerned for New Mexico, responded in kind: "If you can't stand the heat, get out of the kitchen." Even with all the political experience a career in state office had given the elderly Barker, it must have been difficult to read that arguing with him offered "all the thrills which accompany the pounding of sand down a rat hole."[2]

This nasty personal exchange was only one of many, and its tone echoed the dispute within the principal organization of New Mexico hunters, the New Mexico Wildlife Federation. This was the lineal descendant of the New Mexico Game Protective Association, which came together for its first state convention under Aldo Leopold's tutelage in 1916. The 1970s were a far, angry cry from that happier time. The dispute that swept Barker and hundreds of others along erupted over demands by a large group of sport hunters that elk hunting be loosed from the grip of private landowners. For many, the old prerogatives accorded landowners were no longer tolerable. The Class A game parks were as powerful as ever; landowner dominance had even grown in new directions. Because a hunter had to have permission to hunt on any private land, the state usually bypassed public sale of elk licenses, allotting many directly to landowners, who often sold them to favored, wealthy sportsmen from other states. Increasing numbers of sport hunters in Albuquerque now insisted that the state assert its ownership of game as a state commons and take back the power of selling licenses to hunters without interference from landowners. As it was, in the words of one angry hunter, "The Game Dept. works for a handful of ranchers and a couple of thousand out of state clients."[3]

As modest a proposal as it was, for authorities and large landowners this rising sentiment threatened to topple an old tradition of landowner privilege dating back to territorial days. When the protest faction gained control of the New Mexico Wildlife Federation, supporters of the status quo expelled them from the organization. The head office of the National Wildlife Federation had to send an arbitrator, who sided for the most part with the protesters and ordered their reinstatement.[4]

In many ways, this was a conflict Aldo Leopold had foreseen: the day when private landowners and sportsmen fought over game had finally come. Part of the problem lay in simple demographics. The migration of millions to the South and West—the Sunbelt—had begun in earnest. And among those who moved to New Mexico were many hunters, who competed ever more intensively for New Mexico's game. There were 83,000 licensed deer hunters in New Mexico in 1967; by 1974 there were 117,000.[5]

But at the root of this confrontation was the failure of Leopold's "last free hunting grounds" to generate enough game, especially deer, for the rising numbers of New Mexico hunters. Indeed, as New Mexico seemed headed for an abundance in hunters, the wildlife population was falling precipitously. In 1968, state biologists estimated a deer population of 405,000; in 1975, that number had fallen to 276,000.[6] With deer numbers in mysterious and sharp decline all over the state, hunters who had long stalked the public domain and national forests looked across property lines onto the private estates and thought they saw a horn of plenty. They began to demand more equitable rules of access to the hunting grounds, even on private land. Many years before, the state had practically turned over the allocation of elk licenses to landowners. In the early 1970s, the state had similarly ceded antelope licenses to large ranchers who had antelope herds on their holdings. Not surprisingly, a sizable number of sportsmen began to worry that deer hunting would soon fall to the private estate owners unless they took action.[7]

In fact, the decline of deer on private estates was more precipitous than on public lands.[8] And the success of the dissident sportsmen's campaign was very limited: a compromise was reached in which the state took over the official issuing of licenses, but only upon written proof that a landowner permitted the applicant to hunt on his estate.[9] The conflict simmered for some years, but at the mid-1970s reached its ugly peak.

The fight in New Mexico was proof of the perils of undersupply. The heavily federalized state commons in wildlife was meant to secure game abundance, with all its attendant benefits: recreational opportunity for hunters of middling means and a tourist attraction for a cash-hungry state. But scarcity, a condition that reflected changing ecological systems, was disastrous for the political order of this commons. In 1917, Leopold had told the Albuquerque Rotary Club that wildlife was part of a "national environment" that provided "a tremendous social asset" as a source of democratic recreation.[10] The failure of Leopold's "national environment" was far from complete, but in response to the dire shifts in game populations, wildlife management became more heavily federalized than ever. Consequently, the state commons moved ever closer to being predominantly a national commons, in which New Mexico's game was "public goods" for the nation, protected by federal authorities.

This dispute holds clues about how to understand the close relationship between sportsmen, tourism, and federal power. The failure of the national forests to generate perpetual abundance suggests the extent to which any commons—no matter how carefully regulated—is inevitably limited by ecology. To grasp exactly how much sportsmen had invested in securing an abundance of game, and what its failure meant to them, we must look back to the early efforts to organize sport hunters in New Mexico.

In Silver City, sportsmen flocked to Aldo Leopold's banner. Silver City was a mining town, but by 1916 it was also home to many businessmen and wealthy tuberculosis patients, "lungers," who gathered in the area for the dry, supposedly salubrious climate. Many of these people partook of the pervasive belief in fresh air and exercise as a treatment for their malady. Among these upper classes, hunting was a popular recreation. The names of wealthy hunters, at least those affluent enough to take up to three weeks away from town on extended hunting trips, appeared regularly in the city newspapers' society columns.

For these people, recreational hunting took on several meanings, all of them distinctive from the perspective of the region's subsistence and market hunters. A quick survey of local press coverage suggests a number of cultural assumptions about hunting among local elites. Trophies of the hunt were symbols of political and economic power. "Mr. and Mrs. J. D. Hunter and Mr. and Mrs. J. M. Hanks left Monday morning on a two weeks' hunting and camping trip in the Black range," wrote the editors of one local newspaper on October 7, 1913. "They went well equipped with supplies and expect to bring back heads that will ornament any den."[11] Hunting was a re-enactment of arcadian ideals often associated with frontier myth, a romantic counterpoint to the industrial and urban world of Silver City itself. "D. B. Robertson and Charles T. Ross are out on a hunting trip into those sections of the hills where the gentle white deer disport themselves in the sylvan shades and offer venison as a trophy for true aim and steady nerve."[12] To the sportsmen of Silver City, hunting was a ritualistic exercise of power, and the Mogollon hunting grounds were an arena for it, reinforcing their sense of place in the urban world.

Essentially, Silver City hunters were local tourists: living in an urban center and connected to the industrial and commercial apparatus of urban society, they ventured into the mountains for appreciation of an exoticized "nature." Most recreational hunters from Silver City apparently turned north and west for hunting, to the sparsely settled Mogollons and the headwaters of the Gila River.

In September 1913, some of the town's recreational hunters gathered at the Elk's Club to create an organization "to make the protection of wild game and fish more complete and systematic." The result was the formation of the Sportsmen's Association of the Southwest.[13] It is not to discount the sincerity of their appreciation for wildlife to acknowledge that association members hoped for material gain by turning the Mogollons into a tourist attraction. "The object of the organization," relayed the *Silver City Independent,* "is not merely to protect game from needless slaughter, but to take such steps as may be needed to restock the rivers with fish, and such other action as is necessary to make our mountains the attraction they should always be."[14] Just as the penetration of the livestock market into the foothills had changed their numerous local wildlife commons regimes, the rise of tourist hunting marked another shift in market relationships with the mountainous game habitat.

The following month, the editors of the *Silver City Independent* noted, "The number of visitors attracted here by the hunting increases year by year. With proper protection the hunting will last indefinitely, and with our peerless camping out climate should make this a great tourist resort." There was no conflict between the aesthetic appreciation of game and using wild animals to draw tourist dollars, for abundant game and abundant tourism went hand in hand. "The state of Maine has a revenue running close to a million dollars from its hunting alone; and the deer are more plentiful than they were 20 years ago."[15]

In their pursuit of tourist dollars, Silver City wildlife conservationists were allies of state authorities, who had long envisioned a lucrative legal market for wild animals through sale of hunting privileges to tourists. The rise of hunting as a commodity may be read in the changing presentation of the Department of Game and Fish's annual reports. Pennsylvania's reports were slender typewritten pamphlets, with accounting of finances, game conditions, and recommendations for new legislation.

The earliest game condition reports in New Mexico were not unlike these, but by the time it became a state in 1912, New Mexico's game and fish reports had become booster tracts. The report published in that year was a handsomely bound volume with color drawings of fish and photographs of big game, scenic vistas, and trophy bears. Colorful illustrations complemented decidedly purple prose. "It is not exaggerating to say that here in New Mexico, are found attractions to the sportsman, the tourist, the healthseeker, the lover of the picturesque, the sublime, the uplifting, that are unsurpassed in all the world," extolled state game warden Tom Gable.[16]

The men who gathered at the Silver City Elk's Lodge in 1913 to form the new sportsmen's association were an elite group who assumed that state regulation of the hunting grounds meant more game, and more game meant more tourists. Many of them had appeared in the "Hunting Notes" columns of the town newspapers. President of the organization was Miles Burford. Educated in Indianapolis and Chicago, an athlete, the son of a prominent Indianapolis printer, Burford was a "lunger" who had come to Silver City hoping to recuperate.[17] He owned a large ranch north of town, where he kept a pack of hounds for his frequent hunting trips. Other officers of the association included a judge, one of the town's two Buick dealers, and a Princeton graduate who managed a local sanatorium.[18]

The principal mission of the Sportsmen's Association of the Southwest was to "promote the protection of all wild game, fish, and birds of southwestern New Mexico and to assist in the enactment of proper game laws."[19] The number of gun-toting Indians in the hunting grounds was as much an obstacle to the sportsmen's cause as it was to ranchers' peace of mind. In November, two months after the first meeting of the new sportsmen's association, a party comprised of some of Silver City's better-known businessmen embarked on a hunting trip in the Mogollons.[20] The *Silver City Independent* reported an encounter between Louis Jones, one of the party's hunters, and an Indian in a front-page article, "Lone Hunter Meets Indian. Both Were Surprised and the Indian the Worst Skeered—Maybe." To judge from the account, deep fears on both sides characterized such encounters. "To meet alone an armed Indian in the depths of the mountain fastness is the experience which happened to Louis Jones." The Indian "could speak a few words of Spanish and told Mr. Jones that he was alone and that he was hunting." Apparently, white hunters frightened Indians at least as much as they were frightened by them. "Jones told him that there was a large party of white hunters in the vicinity and that it would be safer for him to hunt in some other locality. This he seemed glad to do." As if to apprise readers of the seriousness of the Indian threat, the article included an inventory of the Indian hunter's weaponry and provisions. "The red man carried a Winchester and a six-shooter, also a small bag of grub at his saddle horn." The cryptic remarks of the hunting party suggest that Silver City sportsmen, and probably Silver City residents generally, considered armed Indians a serious threat to life and limb. "The members of the party who went in from Silver City frankly admit that they were just as pleased that Louis Jones instead of one of them stumbled on the Indian." And to judge by clues in the mountains, these encounters were all too likely to be repeated. Although they had only met one Indian hunter, "The party saw numerous Indian signs in the country hunted over."[21]

There was great potential for conflict between these urban sportsmen and the region's rural subsistence-hunting settlers, but in 1913, ranchers and sportsmen stood united against Indian hunters. They also worked to resolve the growing problem of tourist hunters. By December 1915, one Silver City newspaper reported that "the annual flood of hunters" in western Socorro County "has become so large that a great deal of livestock is crippled each year by stray bullets from high power rifles." Local sportsmen were hoping to act as mediating agents between tourists and angry ranchers, planning to distribute safety literature to visiting hunters.[22]

In a sense, sportsmen's enthusiasm for game laws was a crucial lubricant for weak state and federal authorities attempting to stretch thin resources over a large countryside. The rise of sporting associations and their emphasis on tourism in New Mexico guaranteed some degree of support for Leopold's conservationist initiative, for the forester's agenda also included expanding tourism on the national forests. In fact, serving tourists became Leopold's chief economic and political rationale for game protection on the national forests, something the Forest Service had always encouraged, but without commitment of significant resources. In a 1918 article, Leopold argued that just as the national forests served the timber markets, the increasing demand for recreational hunting space constituted a "market" for wildlife. "Given a market, we make game laws (sale contracts) specifying certain license fees (stumpage rates). We may adopt limitations on age and sex of animals to be taken, which are analogous" to rules stipulating which trees were to be cut and which left standing.[23]

Leopold's desire to address the demand for hunting dovetailed neatly with

New Mexico's aspirations for a thriving tourist economy. According to Leopold, the national forests would become a popular destination for the five million American sport hunters increasingly anxious to find open land for hunting. The growing national appetite for hunting meant that it would be "more than ever good business" to manage the forests for game animals, which "will be worth ten times their weight in ordinary venison to the hunter from Iowa or Kansas who will come to seek respite" from posted lands.[24]

In short order, Leopold developed these ideas into a coherent Forest Service policy for protecting game as a national tourist attraction. In 1921, he proposed that the Forest Service exploit the increasing recreational demand for "unspoiled" landscapes by protecting select forests from development. To be called "wilderness areas," these forests would be public hunting lands for sport hunters. They also embodied romantic ideas about "nature" as landscapes practically devoid of human influences: "By 'wilderness' I mean a continuous stretch of country preserved in its natural state, open to lawful hunting and fishing, big enough to absorb a two weeks' pack trip, and kept devoid of roads, artificial trails, cottages, or other works of man."[25]

The first wilderness area Leopold proposed was in southwestern New Mexico, at the headwaters of the Gila River, in the Gila National Forest. (Roughly, this is the area discussed in chapter 3.) Although the region had long been an arena for tense confrontations between ranchers and Indian hunters, to Leopold it seemed a recreational paradise. A half-million-acre tract remote from urban America and without an extensive network of roads, it was ideal for backcountry hunting and fishing.[26] In 1924, the Forest Service acted on Leopold's proposal, setting aside much of the old Gila National Forest as a new "wilderness," the Gila Primitive Area.[27]

At federal, state, and local levels, the new wilderness designation for the Gila filled several important needs. Since the creation of the National Park Service in 1916, Forest Service leaders had worried that the new agency might garner tremendous popular support and power because of the recreational attractions it offered. Now the Forest Service had a coherent recreational policy of its own to counter the growing power of the Park Service.[28]

With the federally owned Gila set aside for recreational use, New Mexico at last had a public hunting grounds for tourist hunters, who provided money for the state economy, gave wildlife conservation an economic rationale and, authorities hoped, broad local appeal.

Although he emphasized its recreational potential for tourists, Leopold was careful to protect the interests of local ranchers on the Gila. Indeed, the new "wilderness" served the needs of the same locals who had been chasing Indians out of the hunting grounds since the early 1900s. In an odd shift from the popular understanding of wilderness as a place where Indians hunted and there were no white settlers, in the Gila there were few if any Indians allowed and many ranchers. In Leopold's (rather forced) logic, visitors would travel long distances to the Gila not just for its wildlife, but also "because of the interest which attaches to cattle grazing operations under frontier conditions." By closing the Gila to both homesteads and road construction, the "wilderness" area would reinforce rancher control of the range, providing stock raisers with the "advantage of freedom from new settlers, and from the hordes of motorists" who would inevitably "invade this region" otherwise.[29]

Being useful to powerful local interests was essential for conservationists; but more and more, New Mexico conservationists turned to national appeals. In little time, they had come to see themselves as advocates for a pool of lands and resources to which a national hunting public had certain bundles of use rights. In the late 1920s, the editors of the *New Mexico Conservationist* exhorted New Mexicans to make their state "the playground of the nation." To their minds, recreational tourism was as valuable and productive an industry as any other, and it was New Mexico's national duty to provide "wild" retreats to an ever more urbanized nation. " 'All work and no play makes Jack a dull boy,' and if our place in the scheme of things is to keep him whetted up for the strife and grind of a workaday world, we contribute no less to the progress of mankind than the state which furnishes it with bread, steel, or cotton."[30] The East and Midwest were "recreation hungry," and the time was at hand when the state could "have a tremendous population of transient vacationists" whose demand for wildlife would increase its value and allow the state to "find its place in the sun by virtue of its recreational possibilities."[31]

This vision of a thriving economy was predicated on abundance of game, which conservationists knew a carefully managed national forest could provide. But for all their care and effort, perpetual bounty on this "last free hunting grounds" proved illusory. Ecological shifts on the Gila and across the entire Southwest brought declines in game, aggravating inherent political tensions in the alliance that built the state commons and sundering the federal-state consensus over deer management in the region.

The first hint of problems to come was in Black Canyon. Abutting the new Gila Primitive Area on its northern border, Black Canyon had been a favorite haunt of Indian hunters in the early twentieth century, subsequently of sportsmen in the teens and twenties. State interests in the area reflected a growing dependence on federal support in matters of game protection. The canyon itself was owned by the federal government, but in 1921, the state exercised its new authority to close federally owned public lands to hunting for game refuge purposes. Although its official name was Black Canyon State Game Refuge, it embodied an intertwining of federal public property and state authority. In closing the area to public hunting, state authorities hoped to create a breeding ground for deer. From here, the animals would continually replenish surrounding forests.

Bad news soon followed; almost immediately, observers began to warn that the deer population was growing too quickly for the habitat. In 1922, a Forest Service range inspector wrote that deer in Black Canyon were overgrazing the range, and in 1926 another inspector made a similar complaint. The irruption was not confined to Black Canyon. Between 1925 and 1929, Forest Service authorities increased their estimates of deer populations on the adjacent Gila National Forest from just under 11,000 to 44,000. In 1930, a Gila forest supervisor noted, "There has apparently been an increase in the number of deer through the entire Forest, but the greatest increase has been in what is known as the Black Canyon section."[32]

The Gila irruption paralleled a better-known irruption of the period, on Arizona's Kaibab Plateau, near the Grand Canyon. By the mid-1920s thousands of deer were starving to death for want of food in the Kaibab's overbrowsed forests. In 1930, one Forest Service game specialist wrote, "Parts of the Black Canyon range are in as bad condition as that of the Kaibab and the area covered by this herd of deer is much larger in extent. Taking the area as a whole there [are] also more deer on the Black Canyon Range than there are on the Kaibab."[33]

What caused the increase of deer on the Kaibab and in southern New Mexico remains something of a mystery, but its most likely explanation lies in the dynamics of plant communities.[34] The dominant species of deer in New Mexico was mule deer. Mule deer are browsers, thriving less on grass than on woody brush. And woody brush was on the increase. Throughout the region, the open country of the grasslands seemed to be closing up; brushy forage plants encroached on open meadows, and the piñon-juniper forests moved down the mountain slopes into what had been more open, brushy areas just ten years before. Aldo Leopold noted this phenomenon in southern Arizona by

1924, where locals complained that brush had "taken the country" in recent years. He saw similar conditions on the Gila in southern New Mexico in 1927, where much former grassland was now covered in fine deer forage plants: oaks, manzanita, mountain mahogany, and deer brush.[35]

Paradoxically, the agent most responsible for producing this brushy, game-rich landscape was domestic stock, notably cattle. Heavy stocking of the range in southern New Mexico in the late nineteenth and early twentieth centuries temporarily decreased the viability of deer habitat in the area, as millions of cattle grazed the land practically clear. Woody plants colonize disturbed soils, and as the region's huge cattle herds converted grassland to barren, trampled dirt they created an ideal seedbed for them. By the early 1920s, mahogany, oak, juniper, and piñon had expanded into the grazed over districts where grass had once predominated.[36] With piñon and juniper providing both cover and mast, and with prime deer forage plants—mountain mahogany, manzanita, and live oak—ever more predominant, the number of mule deer in the region began to grow.[37]

The loss of grass cover had other consequences that compounded the surge in deer numbers. When the grass disappeared, erosion increased dramatically. All over southeastern Arizona and southwestern New Mexico, sheet and gully erosion washed away countless acres of topsoil after the turn of the century, creating widespread anxiety among forest managers.[38] Grass had also been a crucial link between the landscape and the shaping influence of fire. In the years of Apache primacy, grass fires—both natural and the result of human presence—kept woodland at bay and allowed old grassland to regenerate, maintaining the open, parklike landscape. On the national forests, the Forest Service suppressed fire, and the absence of grass as a vehicle for fire made its job easier. But as young woody plants invaded former grasslands, fires no longer killed young brush seedlings and grass roots no longer prevented their expansion.[39] Heavy livestock grazing created a landscape of contradictions, with little to no grass and devastating erosion on the one hand, thickening brush and increasingly abundant deer on the other.

The role of livestock in shaping the Gila Primitive Area suggests some of the deeper paradoxes of Leopold's wilderness concept.[40] Leopold's wilderness was supposed to be preserved in "a natural state." Certainly the Gila's rugged mesas, remote forests, and abundant deer would satisfy middle- and upper-class tourists in search of a "natural" landscape. But if tourists sought out nature as something separate or "preserved" from human or urban influences, then this was an illusory nature. Human handprints were everywhere on the Gila. Perhaps they seemed most obvious in the pronounced erosion that troubled authorities. But some of the very qualities which established the region as "wild" in Leopold's mind—its thick brush and abundant deer—were traces of the livestock market and the bureaucratic machinery of the Forest Service and its fire suppression activities. For all its "natural" characteristics, this "wilderness" was as much a human landscape as the overgrazed Plains of San Augustin to the north.[41]

At odds with the region's past, the romantic idealization of the Gila as "wild" had important effects on its future. Far from preserving the land, the regime which Leopold introduced in the early 1920s profoundly changed the region's ecology. Although Leopold conceived of tourism as a relatively nonconsumptive, low-impact use of the landscape, in reality, tourism's powerful economic force had serious consequences for local culture and landscapes. A comparison of its effects on the one suggests its role in shaping the other. Tourism, in the words of anthropologist Sylvia Rodriguez, is the marketing of local resources "symbolically as well as materially." In 1920s New Mexico, tourism was dependent in part on selling the "exotic" aspects of local life, particularly in Indian communities. Indians were often the focus of tourist promoters, especially the Atchison, Topeka & Santa Fe Railroad and the famed artists of the Taos Circle. Their place in American art history aside, the Taos Circle's chief contribution to the region's economy was the manufacturing of marketable, heavily romanticized images of New Mexico. The AT&SF Railroad sponsored the Taos Circle and bought their paintings for promotional calendars, posters, and other publicity.[42] Led by such luminaries as Ernest Blumenschein and Herbert "Buck" Dunton, the Taos artists specialized in paintings of the "Old West," particularly "traditional" Indian scenes, and in the process did much to shape how the mass public understood Indians. In the vision of "Indianness" the Taos Circle constructed, Indians were inevitably separate from modern life. Tourists came to see them as exotic, mystical, peaceful people, their lifeways seemingly unchanged from days gone by, without historical or social context.[43]

Given their divergence from daily realities, these tourist expectations contributed to a reshaping of local life, and a convincing demonstration that tourism is not necessarily a benign activity. As tourism's preeminent scholar, Dean MacCannell, has written, tourism "promotes the restoration, preservation, and fictional recreation of ethnic attributes."[44] By encouraging Indians to perform select dances, wear "traditional" clothes, and manufacture popular crafts for the tourist market, New Mexico's tourist trade changed the daily activities of many Indians, and thereby encouraged the creation of a heavily constructed, or even "re-constructed" ethnic identity.

Tourist stereotypes of wilderness oddly paralleled popular assumptions about Indians. Just as the tourist's Indians were removed from any historical or social reality, the tourist's wilderness, "preserved" in a "natural state," ostensibly had no history. And, just as tourism promoted the creation or re-creation of ethnic attributes among Indians, it encouraged the preservation of a fictional "natural" environment on the Gila: separate from modern life, insulated from the urban realities of industrial America. Tourists could visit it, and benefit from it, but their presence would not change it. Given the similarity in stereotyping Indians and wilderness, perhaps it is not surprising that Taos Circle artists—including Ernest Blumenschein, Victor Higgins, and Buck Dunton— were ardent sport hunters and conservationists. They became founding members of the early sportsmen's association in Taos and were probably among Leopold's audience when he lectured there in 1916.[45]

In reality, of course, the land was both a product of nature and an artifact. Cattle on the Gila embodied the connection between urban beef markets and the remote landscape. And meeting the expectations of tourists required measures that changed the ecology of the Gila and adjacent areas, however inadvertently transforming its "natural state." Fire suppression was part of the Forest Service effort to "preserve" the nature of the Gila, keeping it usable for wilderness tourism and livestock grazing. Yet, this means of "preserving" the landscape complemented other forces of change and contributed to the expansion of woody plants, which fueled the growth of deer herds.

Conservation and tourism went hand in hand, reshaping local lifeways and landscapes, constructing new ones that appealed to a national tourist market. In a sense, the limitations of conservation and tourism were parallel. Tourism may shape ethnic identities, but it is only one part of the dialogue between local culture and visitors. Indians made their own contributions. While the seemingly peaceful, "traditional" Indians of Laguna Pueblo charmed tourists just off the AT&SF railroad, Laguna hunting parties were in the Mogollons, confounding conservationists and worrying ranchers. The incident of guns drawn between San Felipes and mounted police over doe hides suggests that local realities were far more complex than the manufacturers of tourist images let on.[46]

Indians found meaningful ways of living that did not fit tourist stereotypes, including market hunting and the "slaughter" of deer; local ecosystems developed in ways conservationists could neither predict nor control. By the latter 1920s, the fortunes of deer herds on the Gila, particularly in Black Canyon, took a turn for the worse. Earlier warnings about deer overpopulation seemed to be coming true. In many places, especially Black Canyon, range inspectors reported deer and cattle so numerous that the best browse plants were not regenerating. In 1927, J. Stokley Ligon concluded that the range in Black Canyon and surrounding areas was "badly over-taxed." Deer had browsed local juniper "four or five feet from the ground until little more than the bare branches remain" and were now resorting to a least-preferred food, piñon pine.[47]

Similar reports from Forest Service inspectors brought together the canyon's managing coalition of federal, state, and sportsmen interests. In 1929, representatives from the Game and Fish Department, the U.S. Biological Survey, the Forest Service, and the Silver City Game Protective Association made a joint inspection of the region. They reported that, throughout the area, "the principal browse (oak and mahogany) has been seriously injured." Many individual plants had died, with both piñon and juniper showing heavy use.[48]

To be sure, some of the damage was attributable to cattle, and the Forest Service immediately took steps to reduce their numbers in Black Canyon. Still, authorities were certain that cattle alone could not be responsible for the damage they saw. Over large parts of the range, away from the principal watercourses, there were ample stands of grama grass beneath badly hedged trees and shrubs, a clear indication that deer were overbrowsing to the point of destroying forage plants where even cattle seldom ventured.[49]

The condition of the range continued to deteriorate. In a supplemental survey six months after the initial report, a Forest Service inspector who had been on the 1929 committee stated that Black Canyon's formerly thriving and abundant live oak and mountain mahogany were "practically doomed to destruction unless heavy use of them can be relieved within the next few years."[50] Over the next two years, inspectors remarked on the prevalence of starving deer and the absence of browse. By 1931, desperate deer had begun to eat conifers such as white fir and Douglas fir.[51]

On the assumption that there were too many deer for the range, state and federal authorities took measures to increase hunting in the area. The state opened Black Canyon to hunting in 1929 and 1930; the Forest Service reopened the two major roads into the area, roads which had been closed to protect the ban on motorized transport in the Gila Wilderness. Neither measure had any perceptible effect. In 1930 the Forest Service and the Game and Fish Department joined in an attempt to move deer to other areas of the state where deer populations were low, but the deer refused to enter the corral traps and the experiment was a failure.[52]

As official responses to the deer irruption failed, the interlocking national commons and state commons suffered new political strains. As we have seen, game in New Mexico was a state commons with a strong national presence. It was governed by a hierarchy of interests, beginning with the most extra-local and progressing down to the most local. The most extra-local agency was the U.S. Forest Service. Following it in "extra-localness" was the New Mexico Department of Game and Fish, then the Sportsmen's Association of the Southwest (in Silver City), which took a strong interest in developments in Black Canyon. The most local interests were ranchers.

This complex formation of local and extra-local interests was ideally cooperative, with every level of the hierarchy contributing to major decisions affecting the quality of range. But inherent to such a political structure were various tensions, not the least of which was the inverse relationship of localness to legal power: the further up the hierarchy, the less local was the group and the more authority it had. Ultimately, after all the cooperative meetings were over and everyone's contributions evaluated, the most local interests—ranchers and sportsmen—did not make major management decisions; that was the responsibility of the more extra-local and governmental interests, the state and the U.S. Forest Service.

Already by the late 1920s, changes in Black Canyon's biotic communities' commons were causing these diverse interests to re-evaluate their respective relationships to the land and to one another. The federal government had long paid for predator control in the area. Now the Forest Service urged cessation of anti-predator efforts in order to increase mountain lion and coyote numbers. At least one sportsman excoriated the Forest Service's policy switch on predators in a letter to the editor of a sportsmen's journal, warning that deer "will be exterminated" unless predator control was continued.[53]

Subsequent measures frayed the old alliance of state and national agencies. In 1931, the Forest Service and the Game and Fish Department agreed that reducing the Black Canyon deer herd required a special hunt that allowed each hunter to take two deer, either two bucks or a buck and a doe. This marked the first time that female deer were legally killed since the turn of the century. The

Game Commission opened one hundred square miles of the overpopulated area to deer hunting for twelve days at the end of October.[54]

Although bureaucrats understood Black Canyon's ecology as a deer over-population problem, many sportsmen believed that deer in the area still needed vigorous protection. Making female deer suddenly a category of legitimate prey occasioned a tremor in the conservationist consensus. Particularly vehement in their opposition were members of the Border Game Protection Association in Deming, about fifty miles from Silver City. Members of the association posted large red posters throughout Black Canyon. One was a simple message, which reinforced the pre-eminence of the sportsmen's code over the dictates of policy: "Sportsmen Don't Kill Does." Another poster represented an effort to validate their interpretation of local ecology. Holding forth that a doe hunt would "kill off the breeding stock" and that there was no "oversupply" of deer in Black Canyon, the sportsmen claimed a degree of local knowledge in an effort to undermine the authority of extra-local agencies: "The members of the Border Game Protective association of Deming hunted in this area for years and know it intimately."[55] The Forest Service met these local claims with local action of their own, tearing down the signs and threatening prosecution.[56]

Despite the hostility of many hunters, enough participated to bring in 2,333 of an estimated 3,300 deer in the Black Canyon area.[57] Partly because of this success, the rift between the administrative agencies and many anti–special hunt sportsmen continued to grow. Elliot Barker, who served as head of the Department of Game and Fish from 1931 until 1953, later recalled that the Black Canyon special hunt "brought about the most severe criticism to which the Department has ever been subjected."[58] In an article in an Albuquerque newspaper, one angry critic condemned the hunt as "Not Sport, Slaughter." The critique drew a picture of the special hunt as virtual military assault in which an "army of hunters who invaded the open area and blazed away at everything that had four legs" destroyed the Black Canyon deer herd.[59]

This popular image weakened the formerly strong ties between state authorities and their sport hunting constituents. The Black Canyon hunters of 1931 and the Indians of previous years now shared the same characteristics in the eyes of the hunt critics: both "slaughtered" deer—especially female deer—in military-style forays, with Indians on "raids" and sportsmen "invading" the game country. To hunt critics, the killing of female deer represented an assault on the very ideal of sportsmanship and masculinity. In 1932, a Forest Service memorandum informed the regional office that, as a result of the special hunt, "public sentiment in this region is overwhelmingly opposed to the killing of does" and warned that "the Game Department will have to approach the matter very carefully in the future."[60]

But the situation did not improve. The hunt was supposed to diminish deer numbers temporarily, bringing the animals' demand for browse back into line with what the land could provide. In the middle to long term, deer numbers should have remained plentiful, as soon as the browse had recovered. But this was not to be. Authorities looked on in horror as deer seemed to vanish in the years following the 1931 special hunt. At first, the continuing failure of deer browse plants to grow obscured the fact that deer were also waning. In 1930 and 1931 the Forest Service established several vegetative study exclosures in the Black Canyon area, to isolate different types of grazing (by deer, cattle, or both) and to evaluate their effects on forage reproduction. What the Forest Service found was alarming: on those exclosures where cattle and deer both subsisted, there was a marked decline in mountain mahogany and live oak. Even on those areas open to deer only, there was no recovery of browse, and older plants appeared to be dying. The most disturbing discovery went unexplained: even on exclosures off-limits to all animals—deer and cattle—the percentage of dead plants increased between 1931 and 1936.[61] The special committee report of 1936 concluded that browse plants were declining in part because of "continued use," but also because of drought conditions and other contributing factors such as parasitic infestation of plants, apparently due to reduced vitality.[62] Opponents of the special hunt ignored these findings, claiming that the 1931 hunt was responsible for the absence of deer. To them, the "army" of hunters that had descended on the deer in October had effectively wiped them out.

The sportsmen's perception that there were too few deer now clashed with the interpretations of Forest Service and state authorities, who generally believed that there were still too many deer for the amount of forage in the region.[63] Memoranda between Forest Service officers recommended more either-sex hunting in Black Canyon and on surrounding ranges. But that required state cooperation, and state game authorities were too leery of offending their recreational hunting constituency to order another either-sex hunt.[64]

In the end, the Forest Service decided not to confront the state on the issue. From 1932 to 1942 the buck law prevailed in the vicinity of Black

Canyon. There was an either-sex hunt in 1943 and another in 1954. Until the early 1960s, the Black Canyon area saw very light hunting pressure on does, and, owing to its isolation, only light to moderate pressure on bucks.

This cautious silence between federal and state interests could not mask a broader conflict which Black Canyon had helped to spawn. Because of similar clashes between state and federal agencies elsewhere—on the Kaibab Plateau in Arizona and on the Pisgah Reserve in North Carolina—the federal government began to assume more power over wildlife. More than ever, the state commons in New Mexico—the herds of deer, elk, and other game—seemed to be on the verge of becoming a national commons, managed by federal authorities for out-of-state interests. Traditionally, federal policy ensured that states bore primary authority over game on national forests, and that state game law would be enforced regardless of federal wishes. In the event of a disagreement with federal authorities over wildlife conservation, state authorities would prevail, since they alone could determine the lengths of hunting seasons and the number of hunting licenses to make available. Now the Forest Service sought more power. In 1934, with regulation G-20A, the Forest Service claimed the right to impose its own big game seasons and hunting regulations on any national forest, whenever Forest Service authorities decided that the responsible state govern-ment was not taking appropriate action. The new regulation was a tacit federal assertion that the old political order of the state commons had failed, and that the way of the future would be more centralization of authority in the national government.

State authorities were outraged by this change in federal policy. Forester D. A. Shoemaker, after explaining the new regulation to the New Mexico Game Commission, wrote that state authorities "felt it was more or less of a club being held over the State Game Department by the Forest Service."[65] Generally, state governments across the nation objected to G-20A.[66] Foresters might claim, as Shoemaker did, that the Forest Service would impose its own regulation only after "the failure of cooperative plans," but state authorities believed that the regulation itself signaled the end of a cooperative relationship with the Forest Service. The bitter sentiments of the former allies only empha-sized the extent to which the changing ecology of the commons had under-mined its political order.

The new regulation was rarely applied, and never in New Mexico. But in any case, the political maneuvers of government agencies could not save the deer in Black Canyon. Even with a buck law in force and light hunting, the deer herd of Black Canyon did not recover.[67] By 1960, authorities estimated that both deer and their forage had declined between 65 and 90 percent since 1927.

The Game and Fish Department, searching for an explanation, argued that the decline of forage had been caused by an overpopulation of deer, itself owing to a lack of doe hunting. Game and Fish Department reports in the 1960s noted that "Attempts to establish either-sex hunting in this area are met with extremely strong opposition" because of widespread perceptions that the hunting of does and fawns in 1931 had destroyed the once-abundant deer.[68] Given the political difficulties of reducing deer numbers to reestablish proper ratios between deer and forage, at least one department analyst recommended abandoning deer management plans in the area, and instead beginning a plan to stock the area with elk, which had food preferences that were more suited to the region.[69]

Although the intricate ecosystem dynamics which created this peculiar sequence of boom and bust remain a mystery, there is enough evidence to suggest that sportsmen and state analysts were equally mistaken. Neither over-hunting nor overpopulation accounted for the declining deer population. In the 1920s federal and state efforts to conserve the resources of the forest—game protection, fire suppression, grazing reductions—initiated an expansion of deer browse and subsequently of deer population. Ironically, when continued over a longer period, these same measures led to the end of woody brush expansion and the decline of deer fortunes.

To grasp how this happened, the advance of deer browse in the early 1900s should be considered as a moment in a continuing process of biotic shift. Browse plants like mountain mahogany and live oak aged over time. Before the Forest Service began fire suppression and grazing reductions, fire would have killed many of these plants, and cattle trampling would probably have killed many more. But at the same time, fire and cattle create conditions for new plants to grow. Those plants that provide the best forage for mule deer thrive in conditions of disturbance, such as fire and grazing, which open up the ground for seeds. Without either, the woody growth aged and failed to reproduce. With high deer populations feeding on declining quantities of aging forage plants, the stands of mountain mahogany and live oak began to recede, and so did the deer.[70]

The need to "preserve" the Gila Primitive Area in a "natural state" was only one of the Forest Service policies militating against fire or other disturbances in the landscape. Livestock owners and timber companies were no more in favor of burning the woods than tourists, who viewed the fireless land as preserved, natural. Yet the widely held middle- and upper-class perception that nature was without fire bore some of the responsibility for the changing land and the declining deer. In a curious way, Leopold's wilderness helped create a new landscape of deer scarcity, and consequently the federal-state coalition that the forester worked so hard to build began to fracture.

The perplexing changes in Black Canyon soon became apparent throughout New Mexico. Fire suppression and reductions in livestock grazing everywhere contributed to a less "disturbed" land, in which deer forage plants aged and failed. By the early 1970s, deer were in decline across New Mexico.[71] Sportsmen blamed authorities, and authorities despaired.[72]

Leopold had conceived his "national environment," his "last free hunting grounds of the nation," as a bulwark against the privatization of game. By providing abundant game for the hunting public, the national forests could forestall the expansion of private game parks. When dearth prevailed, sportsmen attacked the private estates, which remained so powerful and entrenched that one expert consultant compared the state's conservation program to Europe's notorious game codes, under which wildlife belonged to the landowner.[73]

With similarities to the European system so entrenched in the state political order, perhaps it was inevitable that conflicts between landowners and hunters in New Mexico echoed European conflicts over game. Sportsmen Concerned for New Mexico demanded that the state abandon its practice of awarding elk and antelope licenses to landowners. Instead, all hunting licenses would be sold by lottery. Landowners could then charge for entry to their estates, or simply deny entry to hunters, but they could no longer dictate who received hunting licenses or what they paid for them. And as the ranchers no longer controlled who received licenses, they would probably have to charge less for "trespass privileges."[74] Although deer were declining on private estates as well as on public lands, many New Mexico hunters demanded that the faltering bounty of the hunting grounds be reallocated because the public had some claim on the private herds of the great estates.

When the New Mexico Wildlife Federation finally split over the matter, the fight had grown so intense that extra-local arbitrators from the National Wildlife Federation intervened. The failure of Leopold's "last free hunting grounds" was not total, and in time deer populations recovered along with ecological conditions. But the bitter debates of these years divided sportsmen from state authorities and seemed proof that the commons in America was potentially as fractious as any in Europe.

The limitations of the commons had been visible in Black Canyon well before the 1960s. Black Canyon and the Gila Primitive Area had long been an arena for opposing systems of hunting and knowing wild animals. Indian hunters who frequented the area knew that if one did not make offerings to the deer spirits, obtaining their permission for the hunt and offering thanks for the kill, the deer would not return. Conservationists figured that state power over the hunting grounds (and the expulsion of Indians from it) guaranteed abundance on the new, centralized state commons.

Yet conservationists could not control the land. Its dynamics remained inscrutable, frustrating the best scientists and confounding the old political alliance of federal, state, and sporting interests. The land had changed, and with it the political realities of living on it, managing it, and hunting its deer. For all the hunters' fierce battles, all the angry shouting over its once-abundant game, Black Canyon was now silent, for the deer were gone.

Blackfeet and Boundaries at
Glacier National Park

In 1928, Glacier National Park Superintendent J. R. Eakin identified a looming crisis on the park's east side. Although the area contained most of the park's scenic wonders, it served "very poorly as a wild life sanctuary or game preserve." The problem was the park boundary, which ran along the shoulders of the mountains. Animals, especially elk, ranged in the park's high altitudes during the summer and fall, but with the onset of snows had to descend. Without refuge from hard winters inside the park, they moved into the sheltered valleys of the neighboring Blackfeet Indian reservation. "Game animals are killed by the Indians as fast as they cross the park boundary into the reservation," wrote Eakin. "Deer and elk cannot increase in this part of the Park until the eastern boundary of the Park is extended eastward to provide winter range for them." Eakin focused his efforts on the acquisition of what he called the "natural" eastern boundary for the park: the Blackfeet Highway, six miles into the reservation. "By extending the park boundary to this line adequate winter range would be provided for all game on the East Side."[1] Proposing a park extension was only one strategy of park superintendents. Where elk could not be controlled, park administrators sought to control Indian hunters through wildlife conservation regimes on the Blackfeet reservation itself. Indeed, in just about any way they could, park officials tried to extend some measure of protection to the elk that crossed the boundary, and in so doing contributed to a profound sense of alienation and deep anger among Blackfeet Indians. For these long-time residents of the region, elk on the reservation were rightfully theirs and besides, park hunting proscriptions were an infuriating betrayal of federal guarantees, constituting a massive "poaching" of Indian wildlife by the U.S. government. For most of the twentieth century, the National Park Service and the Blackfeet contested communal ownership of the animals regardless of their location. In the park and on the reservation, the Park Service claimed them as public goods of the American people; in the park and on the reservation, the Indians claimed them as game of their local commons. Federal officers tried to protect the elk from hunters on the Indian reservation; the Indians frequently ventured across the park line to claim ancient hunting grounds in scenic canyons.

To the first-time visitor, the park's stunning topography masks any hint of social conflict. Glacier National Park is a wedge of land straddling the Rocky Mountains at their northernmost extension in the continental United States. Its defining features are steep mountains and deep valleys, many of them sheltering blue glacial lakes. About two-thirds of the park sits on the west side of the mountains. Starting on the continental divide within the park and moving west, one enters a region of forests and lakes and smaller mountain ranges. Moving east from the continental divide, the journey is very different. The mountains give way briefly to a series of low rolling hills and river valleys, with moderate tree cover. But within ten miles of the mountains, one enters the Great Plains. From the peaks of the mountains, one can see eastward for many miles across the rolling expanse of prairie. Life on the two sides of the mountains was dramatically different; forests and meadows on one side, prairie on the other. In the early twentieth century, local cultures were also dramatically different, with whites living mainly west of the mountains, and Blackfeet Indians immediately to the east.

The plains east of the continental divide had long been the home of the Blackfoot people. "Blackfoot" describes three tribes—the Siksika, the Kaina (or Bloods), and the Pikuni (or Piegans)—that are closely related culturally, and historically allied against the same enemies. In the late nineteenth century, as violent struggles against white expansion came to a close, the Siksika and Kaina lived over the Canadian border, while the Piegans remained in Montana, on their reservation east of the mountains.[2] In a confusing appellation, the Piegans became widely known as "Blackfeet" in Montana. Hereafter, Blackfeet and Piegan will be used interchangeably. Blackfoot, on the other hand, will refer to the three peoples as a whole.

Many of the superintendent's problems in 1928 could be traced to the congressional act that created Glacier National Park in 1910. By that time, authorities could take comfort in the absence of most of the Indians who had once hunted Glacier's mountains. Kutenais, Flatheads, and most Crees had been relocated to reservations miles from the park. The Blackfeet were another matter. The entire eastern boundary of the park, some forty-five miles long, abutted the Blackfeet reservation. Like other national parks, Glacier was conceived as a reserve for monumental scenery, wilderness, and "pristine" nature. More than national forests or any other unit within America's federally managed properties, national parks were expressions of a deep-set, romantic attachment to Nature. They were emblems of America's national heritage, and particularly the heritage of the frontier. Empowering the Secretary of the Interior to formulate regulations "which shall provide for the preservation of the park in a state of nature," the act that created Glacier National Park explicitly called upon administrators to separate the natural from the unnatural, to bound "nature" and hold it apart from corrosive influences, including Indian hunters at its perimeter.[3]

By seeking to move a political boundary—the park border—to match an ecological one—the winter limits of elk range—superintendent Eakin demonstrated a rudimentary understanding of the complexity of his task. There were, in fact, two kinds of boundaries that divided the park and local society to the east. The park line was partly an administrative abstraction that defined the geographic limits of park authority. But beyond that, the park border signified a boundary between reservation landscapes—where locals cut wood, grazed their cattle, and hunted—and national park landscapes ostensibly preserved in a "natural" state. In a sense, the park boundary inscribed on the land an elite, cultural division between the world of people on the outside, and the world of nature on the inside.

Locals, and particularly Indians, contested these boundaries at every level. Blackfeet Indians continually struggled against federal and white encroachment on their hunting grounds, their local commons, and their most consistent target was the Park Service. Killing game as it wandered out of the park or crossing into the park on covert hunting trips, Indians denied Park Service authority and the peculiar barriers that federal authorities tried to erect between local people and local land.

Blackfeet anger with the park was only part of a wider local antipathy to park hunting restrictions; many whites resented the loss of old hunting grounds in the park, too. But in important ways, Blackfeet claims to the land and its bounty were different from those of other local hunters. The Blackfeet were not just "other locals," but among the region's earliest inhabitants. Their use of the hunting territory in Glacier Park went back centuries, pre-dating not just the park but the United States itself. The federal government had in fact promised to protect Blackfeet hunting rights in the area. To a considerable degree, Blackfeet demands for use rights in the park had a moral basis far deeper than the complaints of local whites.

The national commons that engulfed so much of the customary Blackfeet hunting grounds was in many ways more nationalized than other areas discussed in this book. Here, the federal government, especially the Department of the Interior, bore near-total responsibility for resource protection and management. Whereas the Forest Service in New Mexico appealed to strong local constituencies of ranchers and settlers, as well as to national interests, federal authorities at Glacier National Park appealed to a national constituency of tourist vacationers and often ignored local opinion.

In a sense, this national commons was rooted to place. Unlike national forests, where "public goods" like trees and grass for the most part could be separated from the land and sold in large quantities, national parks embodied an attempt to keep land and its "goods" together, "in a state of nature." But in their effort to preserve this landscape for the American people, officials sought to sever the enduring connections between people and the mountains that had made the complex mixture of organisms and geography not just a resource but a *place* in local society. The park was gathered into an abstract quantity of nature, a pool of resources in a category of "national parks" under the director of the Secretary of the Interior. In the minds of its managers, it had no social context, no locality.

While Indians threatened such illusions from outside the park, reminding administrators of the ancient bonds between the Blackfeet and the mountains, the park's own elk, the very embodiment of nature that authorities were sworn to protect, challenged them from within. Migrating elk herds were an annoying reminder of the continuing biological connection between park and reservation, a kind of bridge spanning the park boundary and drawing conservationists and Indians into constant standoffs. The descent of the elk from high mountain slopes each autumn was the prelude to inevitable confrontation and conflict in the valleys below.

This perennial battle was in many ways a consequence of nineteenth-century developments. Prior to the 1850s, the Blackfoot peoples were the dominant power in northwestern Montana and southwestern Alberta, and they had considerable success defending their large communal hunting grounds from outsiders, including the Kalispell, Crow, Sioux, and other hunters in the region. Within the tribe, the hunting grounds was regulated by tribal custom. When the bands of Piegan gathered into a larger community just prior to the summer buffalo hunt, tribal policemen ensured that no independent hunters went after the great herds before the tribal hunt began, lest the animals scatter and the people starve. Punishments varied in proportion to the scale of disturbance caused by the culprit and the supply of meat in the camp, but they could be stern. An impatient hunter who endangered the community hunt could have his weapons smashed, his clothes torn, his whip and rope cut into pieces, and his saddle broken. Had he taken any meat during his foray, the tribe could confiscate it and turn it over to those most in need.[4] Thus, like many other Indian hunting territories, the Blackfeet homeland was a local commons characterized by vigorous exclusion of outsiders and careful monitoring of "insider" behavior.

Between 1850 and 1890, the Piegans, like other Indians of the Great Plains, saw their buffalo hunting economy collapse before waves of white immigration. The Blackfeet negotiated with the newcomers to retain control of their hunting grounds. Meeting with American representatives in 1855, the Piegans agreed to railroad construction through their homeland, and in return received guarantees to all land between the Musselshell River and the Canadian border, from the Rocky Mountains on the west to the mouth of the Milk River on the east. During these years, the buffalo and other game were numerous, and hunting remained the mainstay of their economy. About one hundred miles east of the mountains was a series of low hills, rich in grass and game. The Sweet Grass Hills, as they were known, with their abundant buffalo, elk, and deer, were the principal hunting ground of the Blackfeet.[5]

But in the remaining years of the nineteenth century, a surge of American immigration radically reduced the Blackfeet hunting commons. A gold rush at Fort Benton and a rising cattle market after the Civil War brought thousands of white immigrants to the Blackfeet lands. A series of small skirmishes between Blackfeet and white immigrants, the Blackfeet war, began in 1865 and ended with the U.S. massacre of a peaceful Blackfeet encampment on the Marias River in 1870. Thereafter, the Blackfeet ceased to make even small raids on white settlements.[6]

And the Blackfeet homeland shrank further. A series of unilateral and questionable declarations by the federal government left the Indians with a fraction of their former holdings by the 1880s, a long strip of land between the Canadian boundary and the Missouri River. Even this, white cattlemen eyed with envy. Indeed, many were already using the land for grazing without Indian permission. Officially, no white person could depasture cattle on the reserve without Indian approval, but the Blackfeet had sworn off war against whites and the federal government would not enforce its own treaties against white American citizens. The reservation became virtual free range for the region's cattle barons.[7]

This de facto loss of the older, bounded common area became official in

the late 1880s, when a U.S. deputation arrived among the Blackfeet to negotiate a purchase of "so much of their land as they do not require." Impoverished,
the Indians agreed to sell all but the western end of their reservation, a piece
extending from the continental divide to the mouth of Cut Bank Creek. It was
still a large holding: 1.8 million acres. But the best hunting grounds, to the
south and east, were gone. So was the game. Even in the Sweet Grass Hills, the
elk and buffalo had vanished. Usually, buffalo hunters could take hundreds of
animals in a good year. The last buffalo hunt occurred in 1883, when a group of
hunters returned from the Sweet Grass Hills having killed six buffalo and two
antelope. The treaty of 1888 went down in Blackfeet history as "when we sold
the Sweet Grass Hills."[8]

The Blackfeet homeland suffered one more reduction, with troublesome
consequences for later relationships between the Indians and the Park Service.
In the early 1890s, a tide of white prospectors poured into the western end of
the Blackfeet Reservation, looking for gold rumored to be in the steep mountains. Fearing bloodshed, and once again reluctant to defend boundaries guaranteed to the Indians in earlier treaties, federal authorities elected to buy the
supposed mineral lands from the Indians. A delegation of U.S. commissioners
arrived to negotiate the sale in September 1895. At issue was a tract of land
roughly ten miles wide, from the peaks of the continental divide to the eastern
foothills of the mountains, and sixty miles long, from the Canadian boundary
to the southern edge of the Blackfeet Reservation.[9] Although the Indians demanded $3 million for the land, they were in a poor negotiating position.
Several times during their meetings with the commissioners, the whites re-
minded them that prospectors would continue to enter the reservation and that
keeping them out would be impossible. Walter Clements, one of the negotia-
tors, cautioned the Blackfeet not to refuse the U.S. offer of $1 million for the
land. "It may be a long time before Congress will again make an effort to help
you. It will be very hard to keep people out of the mountains if they think there
is mineral in them."[10] The message was clear: sell, or the lands will be taken
without payment.

Accepting such terms was not easy for the Blackfeet, to whom the mountains were the Backbone-of-the-World, an extraordinarily important place in
cosmology and economy. Exactly how long the Indians had been connected to
this landscape has been a matter of some dispute, but the Blackfeet were
hunting, gathering wood, and worshipping at sacred sites in the mountains
long before whites ever saw them. The most recent scholarship (and a good deal

of the older scholarship) suggests that the Blackfeet have been resident at the foot of the mountains for at least a thousand years.[11] In the pre-horse era, Blackfeet bands hunted buffalo in valleys at the base of the mountains, in coordinated community efforts to drive entire herds over cliffs. After the arrival of the horse, even when Blackfeet men hunted mostly in the Sweet Grass Hills to the east, women, children, and the elderly took to these mountains to gather roots, berries, and timber. The mountains were home to powerful spirits, including Wind-Maker, Cold-Maker, and Thunder. Throughout what became the eastern half of Glacier National Park, Blackfeet men embarked on vision quests as rites of spiritual renewal; on the shores of area lakes, at St. Mary, Waterton, and Two Medicine, they cultivated the sacred tobacco so essential to numerous religious ceremonies, many of which originated in the mountains. So, too, did many sacred objects, including the medicine pipe bundles, repositories of sacred lore and tradition and central to many Blackfeet rituals. The Beaver Bundle, the most sacred bundle and the one used for calling the buffalo each year, originated at a lake on the east side of the Backbone.[12]

When they were not hunting buffalo out on the plains, Blackfeet hunters frequented the valleys. As winter came to the high mountain peaks, elk descended to the canyons and meadows of Cut Bank, Two Medicine, St. Mary, and Red Eagle valleys.[13] St. Mary Valley was the favored hunting grounds of Hugh Monroe, a nineteenth-century white hunter and scout who married into the Blackfeet people. His extended Blackfeet family and their descendants also hunted the area.[14] Although it is impossible to know for certain, it seems likely that during the 1880s and early 1890s, as hunting failed elsewhere, the mountain valleys became more central to Blackfeet economy. Once the Sweet Grass Hills were lost, the tribe looked increasingly to the mountain valleys for traditional foodstuffs and wood, and as an arena for the expression of male hunting prowess and female gathering skills. In 1891, George Bird Grinnell recorded traces of Indian encampments "everywhere" in the St. Mary country, "with rotting lodgepoles, old fireplaces, and piles of bone and hair, showing where game has been cut up and hides dressed."[15]

Indeed, some of these camps were probably in use immediately prior to the 1895 negotiations, at which U.S. commissioners demanded that the Indians sell the mountains, and the Indians made clear their reluctance. As Little Plume observed, "All of the young men who have come here to this treaty were chopping wood in the mountains. . . . If we are hungry we go up to the mountains and get game."[16] Faced with the threat of federal nonenforcement of the 1888 treaty, the Blackfeet knew they could ultimately lose the entire area to white squatters without receiving any compensation. So they made the best deal they could: they accepted $1.5 million for the land, but in a series of special provisions they reserved customary local use rights. White Calf, the leading chief of the Piegan, told the whites, "I want the timber because in the future my children will need it. . . . I would like to have the right to hunt game and fish in the mountains." Big Brave supported him. "I raise my hand to say that we want to hunt game, fish, and cut timber in these mountains."[17]

The final draft of the 1895 agreement transferred the land to the federal government but gave the Blackfeet "the right to go upon any portion" of this land, which became known as "the ceded strip." Under the terms of the treaty, Blackfeet could cut and remove "wood and timber for agency and school purposes, and for their personal uses for houses, fences, and all other domestic purposes." And, in a provision which was to be the focus of a long and difficult struggle, the Indians reserved the right "to hunt upon said lands" and to fish in the streams. After the commission read the agreement aloud, the Indians agreed to the terms and signed. The commissioners left. In 1896, the Senate ratified the treaty, and it became law.

In decades to come, this agreement lay at the center of disputes between Glacier Park authorities and Indian hunters, for what the Indians demanded and what the agreement provided were very different. In a move that under-mined guarantees of Indian rights, at some point in the negotiations the com-missioners had inserted several clauses in the agreement. Although many Black-feet signed the document, the commissioners' additions altered the meaning of the treaty in significant ways. The right to "go upon any portion of the lands" and to hunt and fish there was to continue only so long as the ceded strip "shall remain public lands of the United States." Further, hunting and fishing on the lands was to take place "under and in accordance with the provisions of the game laws of the state of Montana."[18]

The official transcript of the negotiations contains no evidence that the commissioners explained these clauses to the Indians, and the reason for their addition remains a mystery. It is hard to believe the Blackfeet would have accepted the authority of the state Game and Fish Department had they known that the 1893 legislature had banned elk hunting, a common Blackfeet activity in the foothills.[19]

Of greater significance for the future was the public lands clause, which eventually became the centerpiece of government attempts to end all Indian hunting rights in the future park. Once the land became part of a national park, it ceased to be what is legally defined as "public lands." The commissioners probably knew that including the public lands clause would lead to termination of Indian hunting rights at some point. That very summer, in 1895, federal agents to the south were preoccupied with the uproar in Jackson Hole, Wyoming, where Constable Manning's posse had fallen on Ben Senowin and his Bannock companions near Fall River in July. The legal issues of that case revolved around the question of whether Indians retained treaty hunting rights on allotted lands, and whether state game laws superceded treaty rights. By limiting treaty rights to unallotted, "public" lands, and by placing Indian hunting and fishing under the jurisdiction of Montana Fish and Game authorities, the commissioners probably hoped to extinguish Indian hunting and avert another showdown like that at Fall River.[20]

Whatever the outcome, for now the Blackfeet had retained hunting, fishing, and timber harvesting as common use rights in the ceded strip. Reports of gold, silver, and copper brought a brief surge in white immigration, but these proved to be nothing more than rumors. A few boomtowns appeared and as quickly vanished in the mountain valleys, and then the whites left. In 1898 the area became part of the Lewis and Clark Forest Reserve, but the forest rangers apparently did little to interfere with the Indians.

So, for a time, the Blackfeet had in the mountains what they had before: hunting, fishing, and wood-cutting. Walter McClintock, who gained renown as a Blackfeet ethnologist after the turn of the century, visited the area as part of a government forestry expedition a year after the negotiations, in 1896. He met a number of Blackfeet families, including the family of chief White Calf "in tipi camps at the foot of the Rockies," along Cut Bank Creek.[21] The mountains were not a permanent home, but for the families of White Calf, Mad Dog, and a considerable number of others, they were seasonally important for a variety of necessities. The wooded valleys provided a cool respite from summer on the plains, and the transition zone where the prairie and the montane forests met were full of essential foodstuffs, including wild onions, cowparsnip, and especially blue camas. Several Indians kept cattle and horses in corrals near Cut Bank Creek, where thick grass provided good grazing. They also maintained cabins there, but as late as 1896 they lived in tipis most of the time. Stands of lodgepole pine on the mountain slopes were the only source of tipi poles, and in the autumn, McClintock and his hosts went to the mountain forests to load wagons with firewood for the winter.[22]

And the mountains were Blackfeet hunting grounds. Indeed, by 1896, McClintock wrote, the ceded strip was "the heart of their game country," abounding "with elk, deer, antelope, moose, and grizzly bear."[23] McClintock hunted often with a Blackfeet friend that summer. In the autumn he joined a Blackfeet hunting party of men and women bound for the mountains, in pursuit of mountain goats and bighorn sheep, high in the mountain's alpine regions.[24] Between 1896 and the creation of Glacier National Park in 1910, hundreds of Indians hunted, fished, and gathered timber in the ceded strip.[25] Indian uses of the land shaped it. George Grinnell noted that in the area of Swift Current River, at the north end of the park, "Large tracts of forest" had been "burned over by hunting parties of Canadian Indians."[26] Far from the nonhuman "natural" landscape later envisioned by the Park Service, the eastern side of the continental divide was a human environment shaped by generations of Indian residence.

In the eyes of federal authorities, the creation of the park undermined Blackfeet common use rights to the ceded strip more than earlier legal agreements had. To understand how the wood and game of the mountain's eastern ranges moved from local Indian control to the purview of the federal government, we must explore the nexus of local and extra-local interests which created the park. Like the national forests in New Mexico, this national park had a locally conditioned historical context, serving the particular interests of some elite locals at the expense of others, particularly Indians. Moreover, the national park at Glacier also addressed the concerns of national constituencies who allied their cause with local elites to inscribe new boundaries between local people and local game.

The earliest champions of the national park idea in northwestern Montana were not local people, but extra-local elites. Chief among these was the editor of *Forest and Stream,* George Bird Grinnell. An ethnologist who had spent considerable time among the Blackfeet, a leading sport hunter of his day, and a member of the commission that negotiated the 1895 agreement with the Blackfeet, Grinnell had visited the mountains on his frequent hunting trips in the region. By 1891 he had conceived the notion to turn the area into a national park, and soon began extolling its scenic wonders. "Here are cañons deeper and narrower than those of the Yellowstone, mountains higher than those of Yosemite."[27] Grinnell was instrumental in securing the area's designation as a national forest reserve soon after the 1895 negotiations; he lobbied political leaders for the creation of a national park in his beloved "St. Mary Country," and he urged the idea on a broader public in the columns of *Forest and Stream*.[28] At a critical moment in the congressional deliberations, he received strong support from Louis Hill, owner of the Great Northern Railroad, who hoped to duplicate the success of the Northern Pacific Railroad in turning Yellowstone National Park into a popular tourist attraction.[29]

Grinnell's fascination with the Backbone-of-the-World stemmed from his personal conviction that it was an "uninhabited wilderness" and should remain that way. To him it was a landscape harkening to America's frontier past, the mythic conquest of "virgin wilderness" so central to the American saga. With the inclusion of the area in Glacier National Park, Grinnell's vision of the lands became official, and millions of tourists who shared his perceptions eventually visited the new park.[30]

If Glacier was to appear a wilderness, then federal authorities and park advocates had to effectively "dehumanize" the landscape and "delocalize" it, propounding an official case that the Blackfeet did not live in the park and never had. Chief architect of this campaign was none other than Grinnell himself. Although he had followed Blackfeet guides on his own hunting trips in the mountains, Grinnell consistently rejected the notion that Blackfeet had ever utilized the mountains, and subsequent park historians have frequently repeated the mistake.[31]

Such dismissal of customary local uses of the landscape fired local opposition to the park proposal from the beginning, and not just among Indians. For all the enthusiasm of Grinnell and Hill, local settlers west of the continental divide at first resisted the park proposal. There was some local sentiment for protection of the area's scenic beauty—many locals vacationed at Lake McDonald—but there was also widespread opposition to a park because it would forbid further home building, mining, and timber cutting, the region's most lucrative year-round industry.[32] The Kalispell Chamber of Commerce, fearing commercial isolation on the west side of the continental divide, insisted that railroads be given right-of-way across any park. Through legislative channels, these elite local interests ensured that the federal government ultimately accommodated them. The park bill of 1910 may have mandated preservation of the land in a "state of nature," but this "nature" was heavily compromised: the bill allowed harvest of mature timber (at the discretion of the Secretary of the Interior) and granted railroad right-of-way across the park.[33]

If the park legislation allowed some locals to develop and use the land in their preferred ways, there was no such effort to protect the interests of Indians and settlers who used the land for hunting. Designation of the park gave administrators the power to begin making the landscape less human and more "natural." They banned hunting immediately. Park regulations also forbade the cutting of live trees, fishing except with a hook and line, and homesteading. Tourist use rights were carefully delimited so as not to detract from the park's natural beauty; despite its visible human heritage—including its burned-over forests and remains of hunting camps—Glacier National Park was now ostensibly nature without humans.

New boundaries between people and the land occasioned resistance from white and Indian hunters alike. Within the first year of the park's existence, rangers were deployed on the park's northern edge to intercept poachers from Mormon settlements in southwestern Alberta.[34] On the southern boundary, authorities arrested some of the most prominent local citizens for poaching in 1911. Among them was the postmaster from the town of Essex, Thomas Shields, who had numerous confrontations with park authorities.[35] Superintendent Galen reported in 1913 that Shields "is continually a source of trouble for my men, for the reason that during the hunting season his camp is right close to the park line and I have no doubt but that he crosses in at every favorable opportunity."[36] Shields had plenty of company. The Porter and Williamson brothers also bedeviled antipoaching patrols near Essex, where according to the park superintendent in 1914, "some feeling exists against our Park Rangers on account of their enforcing the rules with reference to hunting within the park."[37] On the west side of the park, "some feeling" was also in evidence. When park ranger Doody trailed a pair of hunters inside park boundaries in November 1913, he found them behind a stand of trees—with guns drawn. Threatening to shoot Doody, local hunter Edward J. Warner ordered him to lay his weapon down and leave. The ranger complied.[38]

With white poachers invading the park from north, south, and west, Indians made incursions from the east. In August 1912, the acting superintendent of Glacier complained that rangers frequently met "Indians and half-breeds" carrying rifles in the eastern section of the park. Indians inevitably claimed ignorance about the location of the park boundary; the superintendent doubted that the Indians were so ill-informed as they claimed, and considered cutting a line along the park boundary from the railway to Two Medicine Lake (making the invisible boundary undeniably visible) just to remove this common excuse.[39] Other Indians did not even bother to claim ignorance. In November 1913, authorities arrested L. E. Murphy and Oscar J. Boyd, "two

Indians, or Breeds" who had in their possession two rifles, two horses and tack, and cooking utensils. They were there "for the purpose of hunting which they admitted with full knowledge that they were within Glacier National Park."[40]

Indians and whites seem to have had some common reasons to resist new restrictions on hunting. All came from remote communities, and probably none had any substantial connection with the Kalispell Chamber of Commerce or other local elites who expected to profit from the park's tourist economy. All the region's rural people depended to some degree on ranching, timber cutting, mining, and hunting. In forestalling these activities, the park hindered local means of subsistence.

And, although local economic imperatives drove people across park boundaries for game, the nation's legal boundaries around wildlife were often too permeable to discourage them. Park regulations against hunting did not have the force of law until Congress passed legislation specifically banning hunting there, something Congress did not do until 1914. Because the federal government had no laws regulating hunting or fishing in the park, and because the applicability of state law in the national park was unclear, park authorities and the state of Montana fell to widely publicized bickering over the question of who bore responsibility for game protection in the park. Consequently, some locals claimed that they could not be prosecuted for hunting there. The courts frequently validated their conclusions. When the Essex postmaster, Thomas Shields, was taken before the district court for killing an elk in the park, he succeeded in having the case dismissed.[41] Three Forest Service employees arrested for hunting in the park in fall 1913 were released because the federal government had no jurisdiction over hunting.[42] In white communities, at least before 1914, it was widely known that there were no federal strictures to protect game in the park: newspapers across Montana had given the issue considerable coverage.[43] Consequently, hunting in the park continued to be widespread until at least the mid-1910s. Resentment over the alienation of local hunting prerogatives would have been considerable anyway, but it was exacerbated by the fact that the authorities who prevented hunting did not have clear authority to do so.[44]

Congress finally banned hunting in Glacier Park in August 1914, thereby clearing up most of the confusion.[45] Park rangers could arrest poachers without fear that cases would be dismissed for lack of jurisdiction. Occasional poaching by whites continues to this day, but it ceased to preoccupy park authorities after 1914.

Park authority over white people who lived outside the park was very clear by then, but whites who lived in the park posed a different problem. A number of whites had acquired title to lands west of the continental divide before the creation of the park, and because the Park Service had no game protection authority on private property, landowners within park boundaries retained the right to hunt, fish, and trap on their own parcels.[46] Legally, there was little park authorities could do about this, except to warn against hunting on park land.[47]

Park rangers attempted to close this loophole in the law with a combination of bluffing and harassment. If they hunted legally outside the park boundaries and brought the kill to their own lands inside the park, landowners could be arrested before reaching home and prosecuted for illegal possession of game within the park. "It seems depriving a person of his just rights as a citizen to forbid the having of game and eating of it during the open season," complained a resident of Belton, the present-day town of West Glacier, at the park's west entrance. Apparently, park authorities misled local hunters, claiming that their legal authority to stop hunting extended onto private lands. The same resident of Belton complained that in late 1911, the park superintendent told him that landowners "are not permitted to kill deer on our own claims," and he "insists on trying to prevent our gutting game on our own places during the open season."[48]

These attempts to intimidate landowners nearly backfired. In a 1915 letter explaining what he saw as the major threats to wildlife in the park, state game warden J. L. (Jake) DeHart reported rumors of "a movement on foot by owners of land within the Park to claim the right to kill and take wild game upon their land and permit others to do so." In the game warden's estimation, park authorities faced "a gang of poachers" who "would murderously assault wild game at any time of the year, in or out of season."[49]

The "movement" never materialized, but private landowners continued to challenge the new park boundaries around wildlife. Park resident Tim D. Sullivan inquired of authorities in late 1915, "Is it against the law for a man to hunt on his homestead during the open season in Glacier National Park[?]" Attorneys for the Department of the Interior nervously reached a twofold conclusion. Landowners within parks retained the right to hunt and fish on their own lands—and park authorities would do well to obstruct further local inquiries. "I believe . . . that at this time it is not advisable to directly answer the question submitted," wrote assistant attorney Lange.[50]

Sullivan wrote to his congressman and the state of Montana in an attempt to discover his legal rights, but each time federal officials stonewalled. In 1916 park superintendent Ralston notified park residents that hunting on their own lands was now against the law. "I regret to state that this notification to the settlers created some friction, as several of them boasted openly that they would hunt upon their own private grounds any time in the open season for the killing of game within this state." The settlers' bravado notwithstanding, the superintendent reported that there were no hunting violations by park residents that year.[51]

For years to come, private landholdings within the park were a thorn in the side of park administrators. Hunting was only one of many troublesome activities. Park landowners grazed stock that wandered onto park land, they burned brush and inadvertently started forest fires, and they frequently hunted "park" game when it wandered onto private property.[52]

As intractable a problem as this was, there was a long-term solution. In 1930, the Park Service began buying up private parcels as they came up for sale, thereby turning the lands over to park authority. It was an expensive process, but over time it decreased confrontations between landowners and the park.[53] Over the thirties and forties, park complaints about private landowners declined as federal authorities consolidated control over the park and established a uniform blanket of federal protection for all animals within its external boundaries.

Thus, during the early years of the park, political boundaries were inscribed which increasingly shielded game from local whites. Attempts to draw legal boundaries between Indians and park wildlife were less successful. This was remarkable in itself because authorities seemed to have less trouble convicting Indians than they did whites. Two Blackfeet trappers arrested in the park in 1912 went before a judge who ordered them not to return to the park, sentencing one to sixty days in jail for killing a deer out of season.[54] The prospect of such punishments must have been daunting, but at least some Indians were aware of the 1896 treaty guarantees and demanded that federal authorities abide by them. In November 1915, D. D. LaBreche, a Blackfeet who lived at the park's eastern entrance, requested information regarding Indian hunting rights in the park. LaBreche enclosed a copy of the Treaty of 1896. "It would be an act of injustice to the Indians to take this right [to hunt in the park] from them without recompensations for the privilege which they retained, and never sold to the government."[55]

Not even park superintendents were confident enough of their powers in the ceded strip to dismiss LaBreche's query out of hand. Superintendent Ralston himself requested a legal opinion on Indian hunting rights in the park. According to Ralston, Indians were not only killing animals in the park, but also "have been in the habit of entering the park and driving mountain sheep and other wild animals from the park lands into the abutting Blackfeet Reservation and there killing same."[56]

Responding to Ralston's request, attorneys for the Interior Department took the position that Indian rights to game in the park had ceased when the park was created. Citing the case of the Bannocks in Jackson Hole as precedent, the solicitor at the Interior Department wrote, "Upon dedication of the land to use as a national park, it ceased to be public land of the United States, within the intent and meaning of the term as used in the agreement."[57] Copies of the decision were transmitted to the officer in charge of the Blackfeet Reservation, who had instructions to make its contents known to the Indians.[58] As far as federal authorities were concerned, the boundary between Indians and park game was legally unassailable.

With legal boundaries hardening around it, authorities set about fashioning the park into a landscape that would appeal to tourist conceptions of nature. In a sense, park authorities crossed their own boundary between "human" and "nature" in order to render the land more "natural" for tourist visitors. On the park's west side, Lake McDonald was "visited by thousands of tourists annually and fishing seems to be their largest source of amusement." Fearing that fish would become "nearly extinct" under this pressure, federal authorities were soon stocking the lakes.[59]

At the same time, park administrators and elite recreationists urged oddly unnatural methods to cultivate game for tourists. The park administration considered planting hay in local meadows to feed deer during severe winters, and to bolster local elk populations they imported animals via train from the burgeoning Yellowstone herds to the south.[60] "By establishing salt licks adjacent to some of the points of interest in the park where ample elk range is to be had," park superintendent Galen wrote to his superiors in 1913, "the elk could be seen at most any time by tourists which would greatly add to the attractiveness of the trip" through the park.[61]

Elite recreationists and state authorities agreed that the park should become an elk reserve to lure tourists. Prominent naturalist and writer Ernest Thompson Seton wrote to the Secretary of the Interior in 1916, remarking that after its scenery and climate, "the best asset of the Park is the wild animal life. If this could be increased to such an extent that all tourists could depend on seeing and photographing the larger kinds of animals, it would enormously increase the prestige and public benefit of the Park."[62] Montana's state game warden, Jake DeHart, also urged authorities to feed hay to the elk in winter, "as it has a tendency to make the game gentle and tame, and this is a very interesting matter for tourists."[63]

The degree of Glacier's reliance on tourists suggests some ways in which the national park was distinct from the national forest in New Mexico. In the Southwest, Aldo Leopold was careful to include livestock grazing as a "frontier" attraction in his officially designated "wilderness," and this ensured that the Gila Primitive Area would appeal to local ranchers, a powerful Forest Service constituency. But because they preserved "natural" landscapes, park authorities were oddly responsible for landscapes that had no local residents. Their local constituencies were neighboring business interests who benefited from the park presence, but, in general, national parks relied on extra-local vacationers for their political support. Thus, even more than Leopold's wilderness, Glacier National Park was dedicated to a national constituency of tourists.

Indeed, there was a yawning chasm between the idealized Nature protected by the park and the powerful business interests that supported it. How was the park to find national political support if nobody could live there? How would railroad magnate Louis Hill make money from a "nature" that nobody could use in any of the traditional ways? The answer lay in tourists, who would serve as the bridge between nature and profit. Park supporters like Grinnell and Hill were well aware of the connection. The campaign for America's first national park, at Yellowstone, was largely underwritten by Jay Cooke, financier of the Northern Pacific Railroad. Cooke endorsed the park proposal early on in the campaign, and paid for subsequent lecture tours by promoters of the idea. We need not slight Cooke's appreciation of nature or his devotion to preserving natural wonders to acknowledge that he hoped Yellowstone would become a national tourist attraction like Niagara Falls or Saratoga Springs. When his dream came true, the owners of the only railroad line to the park, Jay Cooke and Company, reaped huge returns.[64]

Glacier National Park was no less a commodity for the tourist trade than Yellowstone. Louis Hill's Great Northern Railroad was a powerful supporter of Glacier Park.[65] The connections between market and the abstract notion of nonhuman "nature" created numerous paradoxes in the national parks. Scenic lands became "park" as a way of protecting their "natural" (i.e., nonhuman) characteristics. The purpose of the park was then to draw thousands of people from unnatural, urban America. A chain of hotel-restaurants in the park would serve the tired, hungry crowds who arrived on the Great Northern; in a sense, the tourist presence "urbanized" the park. The paradox was perhaps best exemplified in the arrival of urban Americans to this reified nature on the railroad, the paragon of industrial, unnatural power.

As if to cap off this series of oddities, the Park Service manipulated the land—a decidedly unnatural practice—hoping for a surge in natural wildlife populations. Protecting game from hunters paralleled park efforts to protect them from predators: the Park Service spent considerable time and money poisoning mountain lions and coyotes in order to save deer and elk, adding strychnine-laced meat to the region's natural systems as a way of making them produce more game.[66] In a way, tourists contributed as much to the *creation* of "wild" landscapes as to their preservation; simply by expecting to see game and making their wishes known to administrators, tourists initiated a management regime that introduced more and more game to park ecosystems.

Park authorities seemed to realize that their attempts to shape the land to tourist tastes would reverberate in ways which were not always beneficial. When Glacier authorities unloaded a shipment of elk from Yellowstone, they "were covered with ticks and it may mean the spreading of them not only to other wild animals but to domestic [stock] as well."[67] Only a few years later, rangers' reports of mountain sheep afflicted with sheep scab set their superiors to finding ways of eliminating the disease, lest domestic sheep populations contract it.[68]

Perhaps the most visible challenge to the goal of a tourist landscape was the migratory habits of wild animals. Even if park authorities could control the incursions of hunters, there was no way of preventing wild animals from venturing out of the park, where they would no longer be under the protective umbrella of park regulations. Elk in particular were prone to wandering. Washington officials complained that the importation of elk from Yellowstone in 1912 "has not been entirely successful," in light of the fact that "elk have been reported killed by hunters west and south of the park in places where they had not been seen for years previous."[69]

Implicit in concerns about ticks, sheep scab, and nomadic elk was the essential fact of the land's uncontrollability—the unwillingness of natural organisms to respond to official necessity. Although their preferred method of dealing with the issue on the park's east side was to seek ways of controlling Indian behavior, the conservationists' problem was one of controlling local ecosystems, or—as superintendent Eakin realized—making the political boundaries of the national park correspond to the ecological limits of elk range.

And, although their legal boundaries around wildlife grew stronger all the time, fortifying the division between Indians and migratory elk on the park's east side became a near obsession of park authorities. Superintendent Chapman reported in June 1912 that there had been hunting of game "out of season probably on the Indian Reserve of game straying eastward from the Park." Federal authorities fought over a solution to the problem. The superintendent of the Blackfeet Reservation was loath to do anything about reservation hunters, maintaining that the Indians had the right to hunt in the park because of the treaty of 1896; others in the Interior Department disagreed.[70]

The presence of the Indians' reservation on the park's eastern boundary, the seasonal movement of elk toward Indian communities, and the tenacity of Indian claims under the treaty guarantees of 1896 made for chronically weak park authority in the area. Successive park administrators responded with proposals to expand the park eastward. In 1913, the park superintendent wrote that deer and elk on the east side "find their fall feeding ground in localities where they can be ruthlessly slaughtered," and therefore he emphatically recommended "an extension of the park on the eastern side, so that the boundary will be about 6 miles further east."[71]

By seeking to reshape Indian prerogatives in hunting, park authorities hoped to alter Indian custom to the point that no Indians killed elk. Only then would there be sufficient game to appeal to tourists, who were the base of the park's support. In other ways, park interests tried to create a kind of Blackfeet ethnicity to draw tourists. Along with the effort to shape a suitably "natural" landscape, this project was well under way by the early 1910s. The Great Northern Railroad, chief sponsor of the park, encouraged tourist travel by hiring Blackfeet dancers at its own Glacier Park Lodge on the park's east side. Like the Indians of New Mexico, whose daily life was increasingly affected by the tourist demand for the sight of mystical, exotic Indians, Blackfeet Indians at the lodge donned "traditional" clothing, told "traditional" tales, and danced "traditional" dances as a way of luring tourists to the "exotic" locale of the park. In a 1912 promotional campaign, railroad president Louis Hill brought ten Blackfeet Indians to New York City, where they camped in tipis on the roof of a downtown hotel. Along with scenic photographs of park landscapes exhibited at the New York Travel and Vacation Show, which was being held at the time, the railroad used the Indians to suggest that Glacier Park was a serene natural setting and the home of a primitive, mystical people. The Park Service might not accept Indian claims to the ceded strip, but on subsequent Indian "camping" trips to Chicago and Minneapolis the railroad billed the Blackfeet as "Glacier Park Indians," and romantic portraits of Blackfeet Indians became a staple of the Great Northern's promotional calendars for decades.[72]

If the Indians at the park lodge could be persuaded to re-create or create anew a Blackfeet culture for tourists, they were less willing to adjust their lives outside of the lodge to suit the Great Northern or park authorities. As far as authorities were concerned, to maintain an image as a "traditional" Indian was one thing, but to be an actual hunter—as Blackfeet men had been for millennia—was quite another. In 1917, park superintendent George Goodwin wrote to the superintendent of the Blackfeet Reservation that at least one of the dancers at the Glacier Park Lodge, a man named Turtle, "had been killing a large number of elk, especially cows and calves, just outside the Park." Park authorities had no jurisdiction over hunters outside the park, so Goodwin requested help from the Office of Indian Affairs. "Any action that you may be able to take, or see fit to take, which will tend to curtail or stop this practice of the Indians of slaughtering the almost tame animals just outside of the Park line . . . will be appreciated by the Director of the [Park] Service and myself."[73]

The Office of Indian Affairs promised to assist, and in December 1917 wrote that "the Superintendent of that reservation reports that the forest guard has looked after this matter and talked with the Indians, advising and instructing them in the premises." To the Superintendent's knowledge, "no elk have been killed on the reservation or in the park since he has been in charge at that agency."[74]

But reports continued to find their way to the Department of the Interior about Indians killing "park" game as the animals crossed onto the reservation in the fall.[75] The state game warden contacted Montana Senator Henry Myers, chairman of the Senate Committee on Public Lands, who in turn requested that some action be taken to stop the Indians. Early in 1918, rangers reported that two Indians were killing elk near the park line. One of these was named "Wallace Munro," possibly a descendant of Hugh Monroe, and, if so, a hunter from a family with a tradition of hunting these valleys since the 1800s.[76] A flurry of correspondence among Washington officials followed. The Secretary of the Interior wrote to Cato Sells, the Commissioner of Indian Affairs, in the mistaken belief that killing elk on the reservation was a crime. "Each winter since I have been in charge of this Department the illegal slaughtering of wild animals in the Blackfeet Indian Reservation has been going on, and I just learned through correspondence submitted to me by Senator Myers that the Indians are killing the animals again this winter." The Secretary continued, "I wish you would give this your personal consideration and devise some means for making the Blackfeet Indians obey the law. There is not much use in our spending thousands of dollars to protect the wild animals in Glacier National Park during the summer only to have them killed as soon as storms drive them over into the Indian reservation."[77]

The Office of Indian Affairs could do little to create a boundary around game after it moved onto the Indian reservation. Some suggested enforcing a closed season for elk on the reservation, creating a game code that could be upheld through the Court of Indian Offenses. The superintendent of the reservation refused this option, claiming that the Indians were aware of their rights to regulate their own hunting, and they knew that whites also killed elk that strayed from the park. Commissioner Sells agreed with the reservation superintendent, as imposing reservation game laws would "cause considerable discussion and resentment among the Indians."[78] Sells had nothing to offer in the way of assistance on the matter, except to say that a new superintendent would shortly take over the reservation, "and the matter will be taken up with him and a special endeavor made to work out some plan to prevent the slaughter of these animals."[79]

In the face of reluctance from the Office of Indian Affairs, the Park Service continued searching for ways to bring those Indian communities bordering the park under their control. At times, the effort was hidden in more innocuous proposals. In 1917, the civil engineer for the park requested legislation "which will permit the regulation of traffic and police jurisdiction . . . similar to that exercised within the boundaries of the Park" over roads on the western edge of the Indian reservation. Particularly pressing was the issue of the Blackfeet Highway, the north-south road that connected the Indian communities at the park's eastern boundary. As the park engineer saw it, the extension of police authority on the highway was "necessary in order that traffic regulations as to speed, irresponsible driving, etc. can be put into effect," and also so that the unpaved roads would not be damaged by heavy loads during the muddy spring months.

This seemingly technical request had large implications, and an ulterior motive. The engineer continued, "In order to best secure this result, it is believed that the eastern boundary of the Park should be moved so as to embrace all portions" of the highway, "or at least that the police jurisdiction of the Park Service should be extended so as to embrace this area, and that this jurisdiction should not only include the regulation of traffic over the roads, but that it should include the enforcement of regulations with respect to the hunting and killing of game within this additional area, for if this is done, it is doubtful if but few, if any, elk or other game would range outside of these limits."[80]

By the late teens, park complaints about hunting on the reservation suggested an impending environmental crisis. In 1920, park superintendent Goodwin made one of the few complaints about illegal hunting that mentioned whites as well as Indians: "The elk, of which there may be perhaps 200 or more in the Park, are not materially increasing in numbers and never will until they are protected on the Blackfeet Indian Reservation and in the forests to the south of the Park, as the hard winters and heavy snows of the Park drive them outside of the Park boundaries where they are killed by white hunters and Indians."[81] In 1921, the Park Service was given police powers over the Blackfeet Highway, but it did little to improve game protection in the region.[82]

In 1923, the Commissioner of Indian Affairs wrote to the superintendent of the Blackfeet Reservation with a novel proposal: if the boundary of the park did not provide winter range for elk, then perhaps it was time to draw boundaries around the animals. Burke invited Campbell's attention "to the fact that the elk referred to are not native to the Blackfeet reservation but that they were captured in the Yellowstone National Park and planted in the Glacier National Park in 1912."[83] As the animals were transplants, the Indians could have no ancestral claim to them and should be advised accordingly.

Whether Campbell attempted the strategy or not, nothing changed. The park superintendent observed in his report for 1923, "The big game is gradually being exterminated on the east side of the park, as deer and elk are driven out on[to] the Reservation by heavy snows, and the Indians may kill at any time."[84] A year later there was a little more optimism, owing to an increase in the park's ranger staff, which "will mean added protection to our game, against poaching by residents of the districts outside the Park boundary, and particularly the Indians of the Blackfeet Indian Reservation."[85]

As the Park Service attempted to expand its claim to game on the east side, Blackfeet countered with formal efforts to erase the Park Service boundary between themselves and the game of the ceded strip. In 1924, Peter Oscar Little Chief drew up a petition to the federal government, demanding that Indians be allowed to hunt in the ceded strip as the 1896 treaty had stipulated, "in accordance with the laws of the state of Montana." The petition was circulated among the Blackfeet, with an explanation of its contents to be read aloud for those who could not read. "We sold to the U.S. Government nothing but rocks only. We still control timber, grass, water, and all big or small game or all the animals living in this mountains. The treaty reads that as long as the mountains stand we got right to hunting and fishing." Little Chief asked Montana Senator Thomas Walsh to introduce legislation specifically allowing Indians to hunt in the ceded strip.[86]

Sent to Indian Affairs officers, the petition was lost in Washington. In 1928 Little Chief sent more letters to Washington inquiring about the fate of the petition. Federal officials disagreed with its substance and replied that Indians had the same right to hunt in the park as whites, meaning no right at all.[87] Before then, Little Chief's demands were echoed in a 1925 lawsuit that the Blackfeet tribe filed against the U.S. government. The suit alleged that the Blackfeet had been deprived of treaty guarantees without their consent or due compensation. The treaties in question included the 1855 treaty, which guaranteed the Blackfeet and Gros Ventre most of the state of Montana as a hunting grounds, and the 1896 treaty, which guaranteed Blackfeet hunting rights in the ceded strip.[88] It took a decade for the U.S. Court of Claims to render a decision.

In the meantime, officials in the 1920s continued to search for ways to force a divide between Indian hunters on the reservation and "park" elk. By the latter 1920s, park officials were resorting to winter feeding to help animals survive the winters.[89] According to superintendent Eakin, animals on the east side of the park received heavy rations of hay each winter "to keep them from seeking winter pasture outside the park."[90] The feeding program provided what the park's ecosystems could not: a winter range under the protection of park game laws. Eakin continued to warn anyone who would listen, including the director of the Park Service in late 1929, "We shall never have much wild life on the east side of the park until the park is extended to include the Blackfeet Highway."[91] Particularly bad was the winter of 1930, when the approximately 125 elk of St. Mary Valley endured a thaw followed by a rapid freeze, which "coated the snow with a crust sufficiently thick to withstand the weight of a full-grown bull elk." Unable to dig through the crust for food, the animals moved out to the reservation, "where they were recklessly slaughtered by the Indians, so that perhaps only 20% survived the winter. It is said that half-breed merchants sold elk meat over the counters."[92]

To avoid a recurrence, park authorities rededicated themselves to winter feeding efforts. In 1933, the Park Service fed twenty-one tons of hay to elk at St. Mary and Two Medicine; in 1934 elk at Red Eagle Meadows and Two Medicine ate twenty-five tons of hay; and in 1935 they received thirty-one tons.[93]

The use of winter feed to control elk migrations strengthened the park's eastern boundary just as Indian efforts to erase it seemed about to succeed. Superintendent Eakin was succeeded by superintendent Scoyen, whose concern about game on the eastern edge of the park led him to study the 1896 treaty with the Blackfeet. What he found alarmed him. Noting that the Indians had reserved rights to enter the ceded strip and to hunt, fish, and cut wood there, Scoyen used understated language in a letter to the director of the Park Service. In view of the Indian claims, "just what legal right have we to prevent the Indians from cutting timber on the park land or from hunting or fishing in the area which they sold to the Government under this agreement?" Only twenty years before, authorities had avoided answering questions about landowner rights in the park as a way of bluffing locals into accepting federal restrictions on hunting. Scoyen suggested a similar pattern of evasion and bluff on this issue. "It may be that it will not be to the best interests of the Park Service to have a decision made on these matters."[94]

Scoyen's concern was probably heightened by a rumor of impending trouble. In 1932, Peter Oscar Little Chief—the same man who circulated the 1924 petition, and who followed up the petition with letters to officials in 1928— sent a new letter to the park superintendent demanding restitution of Indian hunting, fishing, and timber rights in the ceded strip. The letter (which has disappeared from Glacier National Park archives) rattled Scoyen, who forwarded a copy to the Park Service director's office in Washington. The superintendent was resigned to inevitable confrontation. "There is every indication that one of these days, we will perhaps have to go through court on this matter and we may as well decide where we stand now as later. It would appear that legally we have no right to prevent the Indians from cutting wood and timber in the park for agency and school purposes and for their personal uses; for houses, fences, and all domestic purposes."[95]

Scoyen went on to describe the park's troubled history with the Blackfeet. To his mind, the 1896 provisions reserving Indian use rights in the ceded strip were "the heart of the so-called Blackfeet problem." The park's refusal to accept the 1896 treaty had led to constant Indian resistance to any kind of Park Service control over the Blackfeet Highway. "The resentment which the Indians have over being arrested for hunting in the park and not being allowed to cut timber explains why every proposition we have ever made them regarding scenic easements along the Blackfeet Highway has been so promptly and unanimously rejected."[96]

Scoyen's worries about the park's legal vulnerability to Indian claims soon evaporated. As a result of Blackfeet demands, especially those of Peter Oscar Little Chief, the Department of the Interior requested a legal opinion from its solicitor. The solicitor, Finney, reporting in 1932, agreed with the opinion of an earlier solicitor, West, in 1915: Indian rights to the ceded strip ended when it ceased to be public land, and therefore the legal Blackfeet commons in the ceded strip was extinguished when it became national park property in 1910.[97]

The U.S. Court of Claims reinforced the solicitor's opinion in 1935. Finally rendering a judgment on the case filed ten years before, the court decided that in fact the U.S. had violated the 1855 treaty, but that the 1896 treaty rights were not relevant. Despite witness testimony that hundreds of Blackfeet had hunted, fished, and gathered wood in the ceded strip after 1895, the court concluded that prior to the park's creation, "the Indians of the Blackfeet Reservation did not exercise to any appreciable extent the rights reserved" in 1895 "to hunt and fish in and remove timber from the land ceded in the agreement." Therefore, "such rights were authoritatively terminated by the limitations of the act" creating the park in 1910.[98]

The solicitor's opinion of 1932 and the court's decision of 1935 may have temporarily bolstered the Park Service's confidence in the eastern boundary, but it could do nothing to increase its ability to control park ecosystems. While the solicitor and the judges on the Court of Claims cogitated on the law, the park's ecological systems began to show curious changes that strained its political boundaries in new ways.

For years, park authorities had reached across their own rhetorical divide between "nature" and human society, reordering the land in ways that would increase wildlife populations and thereby appeal to tourists. To their minds, winter feeding of elk ensured elk abundance, creating a more "natural" landscape for tourist enjoyment. By the mid-1930s, nature seemed to be reaching into the human world to reorder the politics of wildlife conservation: suddenly, worries about game protection seemed out of place, for there were too many deer in the park. At one time forced out of the park to find food in the winter, large numbers of deer now remained in the park throughout the year to get the winter feed that park officials were providing. By 1937, observers had been noticing localized overbrowsing for several years. In effect, "artificial" winter feeding had created a new "nature," in which deer were far more numerous than before. This threatened to defoliate sections of the park, as hungry deer descended on jack pine, cedar, and shrubs. The solution, as summed up in the superintendent's 1937 report, was to cease the "unnatural" practice of providing the deer with hay. He hoped that "natural controls" (by which he meant starvation) would "eventually reduce the number of deer in these [overbrowsed] areas to the carrying capacity of the range."[99]

Park policies sought to ensure that the mark of humanity was absent in the "wild" which the park represented, but in fact there was no way to facilitate game survival without leaving human handprints on park "nature." In killing predators, and especially in helping game survive the winter with extra nutrients, authorities helped create a landscape of abundant wildlife. Even with the elimination of these more interventionist measures (park authorities stopped killing predators in 1931, even before they stopped winter feeding), park ecosystems were protected from human uses—like hunting—which had a long tenure as components of natural systems in the region.[100] Despite the dichotomy that park authorities drew between local uses of the land and the pristine nature of this national commons, human beings were no less a part of natural systems in the region than they had ever been. The issue at Glacier was not whether there would be any human influence on the landscape; rather, the struggle was over who would have the privilege of shaping it: Indians, who manipulated it for local use, or federal scientists and park rangers, who constructed a "pristine nature" for extra-local vacationers.

Deer irruptions were only the beginning. A new challenge to the troubled relations between the park and the Blackfeet began not in the contested eastern border of the park, but in the south. There, near the town of Essex, where the poacher postmaster, Thomas Shields, so frustrated park rangers in the early years, a strange new threat had emerged. "The consistent increase of elk in the park," wrote superintendent Scoyen, "is rapidly resulting in an acute winter range problem."[101] Where authorities had for years dedicated themselves to shoring up the park boundary, the suddenly burgeoning elk herds of the park soon brought about a new appreciation for permeable boundaries, the migratory habits of elk, and Indian hunters.

6
Erasing Boundaries,
Saving the Range

In 1927, Lucy A. Sweeney of Painseville, Ohio, wrote George Bird Grinnell a letter of thanks for his long efforts at creating Glacier National Park. Sweeney had visited the park that summer after her mother had died, "and the loneliness and the emptiness in the home seemed to increase, so I ran away from it for a time." For Americans like Lucy Sweeney, the "wilderness" of Glacier Park was both retreat from and tonic for urban ills. "The solitude and vastness of the place seemed to make me saner and the peace and beauty I found everywhere has helped to bring me to a better frame of mind." Indeed, the grand drama of nature in the mountains suggested to Sweeney the triviality of her own problems. "While we are struggling with our burdens in our little places, the great forces there are carrying on ceaselessly. It all seems like a great dream. . . . I want you to know I appreciate your efforts in making available to the public so lovely a place. I shall always see your mountain and your glacier."[1]

George Bird Grinnell had first come to know these mountains as a hunter, and in a sense Glacier National Park, like much of American conservation generally, was a hunter's legacy.[2] In years to come, that legacy became available to a broader segment of American society, as the expansion of the interstate highway system and the wide availability of the automobile brought millions to the park. The preservation for public enjoyment of "Grinnell's Glacier" seemed testament to the benevolence of the old sport hunter.

For the Blackfeet, the mountains and the elk were a different kind of legacy. To these hunters and children of hunters, the elk and the ceded strip were more a local lifeway—a local inheritance—than they were national property. In a sense, the conflict on the park's eastern border was a collision between heirs: those who claimed the park as part of the last will and testament of George Bird Grinnell, and those who claimed the landscape by virtue of the ancestors who had hunted it, and lived it, before.

For these mutually hostile interests, the changing ecology at Glacier could mean very different things. For local Indians, growing elk herds did little to soothe tribal anger over the alienation of their common hunting area in the mountains. But for federal authorities, surging elk populations entailed rethinking elk protection policies and the official relationship with the Blackfeet. To a degree, the shifting ecology of these valleys blurred Park Service boundaries between local people and "park" game; as the ecology of the boundary changed, the boundary itself seemed to change also. Inscribing the boundary was a federal project, and the previous chapter was devoted largely to federal perspectives on it. Understanding the boundary's transformation and the meaning of that development for Indian hunters is the focus of this chapter. It requires a careful examination of ecological dynamics and Indian society and culture along the park's eastern border.

For the Blackfeet, elk had been a vital connection to the mountains for as long as time. In the old days, there had been no strict regulation of elk hunting like there was of the annual buffalo hunts. Most elk hunts probably occurred in the fall, when the herds ventured out of the mountains for winter range and found Blackfeet still in the mountain valleys, gathering and hunting until the first big snow sent them to winter camps along the Marias River. Besides food and clothing there were other reasons to hunt elk. Elk teeth were valuable; before the reservation days, they decorated the most valuable women's dresses, and one hundred of them could buy a good horse. Elk were powerful spirit animals, and pictures of them on the elk tipi called their supernatural power to heal the sick. Most people on the reservation probably knew about the giant bull elk who lived in the lake at the head of Cutbank Creek; when he rose and flapped his ears, the wind blew strong down the canyon, and when he sank beneath the waves, the wind stopped.[3] Sitting around lodge fires in the long winters, probably every Blackfeet had heard some version of the story in which Napi, the Creator of the Blackfeet and a supreme immoral trickster, tried to kill all the elk in the world out of sheer greed and malice. He tricked them into jumping over a cliff, and there would be no elk today were it not for the one pregnant cow who escaped and ensured the survival of the animals.[4]

Elk had another attribute by the 1940s: they were a focal point for park authorities' hostility to the Blackfeet. Indians lived in small communities across the reservation. But particularly troublesome for officials were the largely Indian settlements of Babb and St. Mary, just east of the park boundary, about four miles apart, at the mouth of the St. Mary and Red Eagle valleys. Each time park authorities proposed extending the eastern boundary of the park to the Blackfeet Highway, they were arguing for the inclusion of Babb and St. Mary in the park, presumably in order to remove their people—and their troublesome hunters—from the vicinity of "park" elk.

Blackfeet traditions were not the only ones in these two communities. Although they were part of the Blackfeet Reservation, Babb and St. Mary were populated by a diverse mix of ethnic groups whose connections to one another did much to shape their relations to government authorities. Blackfeet were dominant in these towns, but Crees were also present in considerable numbers. In the late nineteenth century, Cree and Chippewa Indians who participated in Canada's *métis* uprising of 1885 fled to the United States when the rebellion failed. Fearing persecution by the Canadian government, they crossed the international border and moved in small groups onto their old hunting grounds, near the present-day city of Havre, Montana, and to the mountains in what would become the northern and eastern sections of Glacier National Park.[5]

The arrival of Chippewas and Crees occasioned considerable ethnic tension. Blackfeet and Cree people, once allies, had become enemies in the course of the nineteenth century, and many whites viewed the refugees with misgiving. Ranchers, looking to secure their hold on the range, and railroad interests, who feared that Indian settlements would block railroad expansion in Montana, were continually demanding that the government remove these "landless" Indians. In 1916, the U.S. government finally created a Chippewa reservation on the lands of old Fort Assiniboine near Havre, far to the east of the Blackfeet reservation. By this time many Chippewas and Crees had intermarried. A large number of Crees joined the Chippewas at the new reservation, named for the Chippewa leader Stone Child, or—as he was known in the white press— "Rocky Boy."[6]

The Chippewa-Cree resettlement at the Rocky Boy Reservation eased some long-standing tensions, but at the western edge of the Blackfeet Reservation there remained a small Cree settlement in the town of Babb. Babb had been a trading post and inter-tribal gathering place since at least the 1870s.[7] With this Cree settlement, the community retained some of its multi-ethnic heritage into the twentieth century. The Crees having intermarried with Blackfeet families to some degree, or having settled and not wanting to move on again, a few Cree families declined the offer to live at the new reservation. There were Blackfeet in Babb also, but a distinguishing characteristic of the settlement in the twentieth century was its sizable proportion of Cree inhabitants. Among the Blackfeet, Babb became widely known as "Cree Town."[8]

The history of St. Mary is a little more obscure. A hunting area for various people in the nineteenth century, it became a mining settlement in the brief period at the turn of the century before the mineral rush failed.[9] In 1901, the U.S. Census recorded 111 residents in the township of St. Mary. Babb did not appear separately.[10] In 1930, the town of Babb had a population of 478.[11] Residents born in Babb in the 1920s recall that the area's population has not significantly changed in their lifetimes.[12]

People of other ethnic groups also established themselves in the area. Kaina or Blood Indians from Canada and white immigrants from Europe and the East married Blackfeet women and settled on the reservation's western boundary, especially at Babb and St. Mary. When the park was created, along its eastern boundary was a complex ethnic mixture, in which white, Blackfeet, Cree, Chippewa, and perhaps a few others developed complicated relationships which at once welded them into communities and established strong class distinctions among them. Among families of mixed white-Blackfeet parentage, such as the Pendergrasses and the Hinkles, there were connections both to the white world of commercial agriculture and stock raising and to Blackfeet tribal membership and its concomitant benefits. White men who married Indian women, along with the sons of these unions, were known to settle on the reservation and become the tribe's most successful cattlemen. Across the reservation and in St. Mary and Babb, "mixed-bloods" dominated the tribe's wealthiest strata after 1900.[13]

Blackfeet without strong connections to the white world were usually poorer than "mixed-blood" Blackfeet, but at least in Babb and St. Mary they were not the poorest. To be a Blackfeet tribal member, by birth or marriage, entitled one to certain rights on the reservation, like hunting at any time of the year. Without large kin networks on the reservation, Babb's Cree people were generally among the poorest of the reservation poor. As late as the 1950s, Babb residents lived in homes made from studs and tarpaper, with no insulation. They covered knotholes in walls with tin can lids and drew water from a town pump near the tiny schoolhouse. There was no indoor plumbing.[14]

Life was hardest in winter. Even as late as the 1950s, winter storms could block reservation roads for days on end, isolating Babb and St. Mary until state snowplows could clear the roads. Residents with cars then followed the plows into Browning, forty miles away, to buy supplies. Afterward, this caravan of snowplow and cars returned to the small, ramshackle town. Buying supplies, of course, required money, which during the long winters was hard to come by. Park road crews and haying jobs on local ranches paid in the summer, but in winter there was almost no work.[15]

The complicated patterns of origins in Babb, where so many people came from Cree, white, and Blackfeet ancestors, meant that ethnic identity was constantly negotiated and renegotiated. As in other places, ethnic interaction extended from intermarriage and the joint raising of families to fistfights. Cree children often fought with Blackfeet children; Babb adults, Blackfeet and Cree, acquired a reputation for drinking, lawlessness, and brawling which gave rise to their current collective nickname on the reservation, half derisive and half affectionate: the Babbylonians. Local lore about them is perhaps best summed up in a local homily, "You're not a real Babbylonian till you're missing a front tooth."[16]

According to local residents, every family in Babb had an illegal hunter or two. Blackfeet hunters were not averse to stalking game in the park; Cree hunters were hunting illegally whether on the park side of the line or not, so park regulations made little impact in and of themselves. The park boundary ran in a north-south line just a few miles west of Babb, touching the western edge of St. Mary. Frequently, herds of elk passed between the two towns on their way to winter range. Mountain goats, Rocky Mountain sheep, and especially mule deer were also forced to lower elevations by the snow.[17] As rural, cash-poor communities, St. Mary and Babb depended on the availability of game at all times of the year, but especially in winter.[18] Elk and deer meat held off famine, and hides provided leather and clothing.

For Indian hunters on the reservation, events at the southern end of the park, where authorities noticed a rapid increase in elk populations in the 1930s, probably seemed remote. Locals rarely hunted so far away. Still, what happened in the southern districts of the park was a sign of things to come on the park's troubled east side. Locals remember there being many mule deer in the region in the twenties and thirties, and always some elk (though not necessarily as many as there are now).[19] How many elk there were in St. Mary Valley by 1910 is hard to say. Most accounts claim there were few. When Vernon and Florence Bailey surveyed the region in 1917, they recorded that "In all, there may be 50 elk along the east slope of the park, and it is doubtful if they are holding their own against the severe climate and the necessity of coming down to the Indian reservation in the winter, where they are unprotected."[20] The Baileys' survey was impressionistic, and their reckoning of fifty elk for the entire eastern side of the park is possibly an underestimate. George Bird Grinnell recorded estimates of 1,100 elk in the southeastern section of the park in 1911. However many elk there were, park records reveal that Indians on the reservation were hunting elk as they came out of the park in the fall of 1917.[21]

In the early 1930s, the elk in this area began to increase as a result of protection from hunters and park winter feeding. Local hunters took full advantage, especially in the fall. Sometimes the elk walked onto frozen Lower St. Mary Lake, and hunters took to the ice in snowshoes. Practically every family in the area had "elk dogs," mongrels that would circle snowbound elk, harassing them and keeping them in one place until hunters could despatch the animals with a .30-.30 or a .270 Winchester.[22]

Winter, when hides were thick and snow made for good tracking, was a customary hunting season, and Indians came from deep within the reservation to take part. Jess Monroe, a descendant of Hugh Monroe who lived on the reservation, took his entire family into the mountains just south of the park each winter. These mountains were also part of the ceded strip, but having been left out of the park they were subject to state game laws. Monroe and his family made their way south through mountain valleys during January, hunting deer, elk, and other game, ignoring state regulations on closed seasons and bag limits. When their packs were full of meat, they took them out on horseback to waiting family members. In trips that lasted for weeks at a time, Monroe and his hunting family acquired a supply of meat and hides that lasted for months and supported not only the immediate family but poor relatives and friends.[23]

In Babb and St. Mary, local people could take similar amounts of game without leaving on extended trips. During the annual elk migrations, it was not uncommon for hunters to kill several elk in one day. Family and friends helped with dressing the carcasses, and meat was shared among the needy. When Clarence Wagner killed an elk one year, his mother made him a coat from the skin. Lined with silk, the coat protected him from the fiercest winter blast.[24]

Beyond subsistence, game meat—especially elk meat—was an important lubricant of the local economy. Local ranchers, Blackfeet and white, frequently paid Cree hay gatherers with venison. Many traded elk meat to Hutterites in the town of Cut Bank, exchanging wild animal protein for vegetables when their own gardens or pantries ran low. And there was always a cash market for game and fish. Using gill nets, local men and boys could catch dozens of whitefish in a single day. Selling them to kitchens at the park lodges, they found their own way of accessing the tourist market. Other meat could also be sold to passing tourists, and some people sold "traditional" buckskin shirts and moc-casins to tourists and collectors who passed through on their way to or from the park.[25]

Hunting and fishing were not the only skills that local Indians could market. One tourist noted in 1939, "Just outside the St. Mary entrance to the park is an Art and Craft Shop where tourists may buy bona fide Indian hand-icraft. It is run by the Indians themselves and beaded jackets, bags, baskets, dolls, pictures and belts made by the Blackfeet are on sale there."[26] The version of Blackfeet ethnicity on view at Glacier Park Hotel had its Indian-owned counterpart here. Indians in the hotel were paid cash for their services, but the benefit of their efforts went largely to the Great Northern Railroad. Here, in St. Mary, they had a counterclaim to the profits generated from sales of this "reconstructed ethnicity."

Hunting, fishing, and handcrafts all illustrated the complex connections between the park and the reservation, between the national commons and the local commons. Netting fish and selling them was illegal in the state of Mon-tana, as was selling game meat. Indians could pursue these market channels because they remained legal on the tribal hunting commons, the reservation, and because the park provided the Indians with a tourist market for the com-modity of both fish and handcrafts. On the western edge of the reservation, local livelihood fused local commons and national commons in a union shaped to a considerable degree by local prerogative.

Hunting in the park was a way of turning the national commons to local use in a more direct way than selling goods to park tourists. Elk meat acquired there was just as nutritious and marketable as that acquired on the reservation, and it had a political significance. To be Blackfeet in Babb and St. Mary was to claim access to park game. Denying legitimacy of federal authorities in the ceded strip was an essential component of local identity, and, as such, of Blackfeet ethnicity.

Men who were raised on the border of the park grew up with stories of friends and relatives who routinely duped park rangers by decoying them. Entering the park in pairs or small groups, locals divided in two. After giving the first party time to reach a herd of elk or deer some distance away, the second party fired several shots into the air to divert the rangers. The other then made a kill quickly, after the rangers had already set out for the decoy party, who by this time was on its way out of the park. Before a ranger could reach the real hunters, they—and their kill—were gone. Night hunting trips were customary; if the local ranger went to East Glacier on a shopping trip or to meet with park supervisors, quite often a local party would pick up rifles and head into the park.[27]

Of course, ideologies of poaching as resistance were not restricted to these two small towns on the Indian reservation. Indeed, the resentment of federal authority in the Glacier region has sometimes catapulted local poachers to heroic status in northwestern Montana. Just as long-term struggles over land have created myths of bandit heroes in diverse regions of the world, the tensions between local residents and federal authority contributed to the notoriety of Joe Cosley at Glacier National Park. The story of Joe Cosley is superficially very simple: he was a poacher who evaded park authorities continually until 1929, when he was captured. Yet his story also illustrates a mixture of resistance and accommodation to the park which Joe Cosley and other locals, especially Indians, wove into their lives in intricate ways.

Cosley was born in the 1870s, in Canada. According to legend, he was at least part Chippewa. By his early twenties, he was a ranger for the U.S. Forest Service in Montana, and he became a park ranger at Glacier in 1910. Shortly thereafter, he was dismissed for trapping beaver in the park. At some point before or after his dismissal from the park, he was decorated for his service as a sniper for the Canadian military in World War I.

He continued to trap for his livelihood after the war. One of his favorite trapping grounds was in Glacier National Park, in the Belly River country, where he was finally captured by park ranger Joe Heimes in May 1929. Taken to the town of Belton (now West Glacier), he was fined $100, and his traps and other equipment were confiscated. Thereafter he steered clear of the park. A careful dresser with shoulder-length hair, Cosley cut a dashing figure on his brief visits to local communities. He cultivated romantic stories about himself, and this air of mystery made even more intriguing this solitary figure who was renowned as a kind and elegant man. Alone and afflicted with scurvy, he died in a Canadian cabin in the winter of 1944.[28]

Cosley was well known in northwestern Montana and southwestern Alberta, and how his story has been told in the region would make for a worthwhile study in folklore.[29] What attracts attention for this book is Cosley's relationship with Babb, a town that has claimed him practically as one of its own. George Hinkle, a long-time Babb resident, often told stories about hunting in the park with Cosley.[30] Today, the oldest generation of Babb residents remembers Cosley well. He often hiked to Babb from the Belly River country, ten rugged miles away. In town he bought his supplies, and occasionally he stayed with local friends. His hiking ability, in Babb and beyond, was legendary. One man tells about the time Cosley hiked to town for supplies and was subsequently invited to a dance. Rather than stay for the dance attired as he was, he hiked back to his camp to dress in his best clothes, then returned to town.[31] The ranger who arrested him in 1929 recalled that Cosley took steps "like Bigfoot."[32] One local man, remembering Cosley's penchant for hiking vast distances over a short time, comments, "I'd hate to walk with him."[33]

Cosley's life embodied many of the paradoxes of local life for Babb residents. Like the ideal of local living then and now, he lived in remote areas but was sociable, to the point of dressing up for special occasions. His sociability was one aspect of his value to the community; he was a good neighbor. One man recalls that Cosley frequently brought gifts of elk meat on his visits to town.[34] Cosley worked for the park and poached in the park; local people depended on the park, but they used it as much as possible in alternative ways that expressed local power. And Cosley was successful. As one local man summarizes Cosley's life, "He never went hungry, that boy."[35]

To be successful in Babb was to be like Joe Cosley. And a crucial component of Cosley's identity may have reinforced his connection with locals there: Cosley was at least part-Indian. Whether he called himself an Indian, others have since. Park ranger Joe Heimes, recalling the time he arrested Joe Cosley, calls him an "Indian poacher."[36] At least one Babb man positively identifies Cosley as "a Chippewa Indian."[37] Indian, poacher, and living off the bounty of local land: this was Joe Cosley. This was also almost every other local man. Whether one knew him personally or not, in a sense, to live in Babb was to know Joe Cosley. The power of such connections was expressed in the tales of Teddy Burns, a Blackfeet man, lifelong resident of Babb, and bête noire of the federal rangers whom he routinely evaded on his hunting trips into the park. Burns's hunting career began as a boy when he received a gift of an old rifle from Joe Cosley himself.[38]

Cosley in part came to symbolize an ideology of resistance to park authority which had developed alongside the park itself. As park authorities maneuvered to gain control over Indian hunters on the park's eastern boundary, Indians hunters challenged park authority through hunting in the ceded strip and refusing to limit their take of elk and other game on the reservation proper. Park authorities assumed that the elk of the east side were constantly in danger of extinction because of Indian hunters. For their part, Indian hunters' experience told them that authorities would attempt to deny them the elk, whether they hunted the animals legally on the reservation or illegally within the park.

Changes in local ecosystems challenged this oppositional relationship in new ways, beginning in the early 1940s. During these years, it became clear that the surge in elk populations at the southern end of the park was only part of a wider phenomenon. The tone of park authorities' reports shifted markedly as a result, from concern about low animal numbers to worry about too many. At first, this dynamic was restricted to the park's western and southern sections. "Elk seem to be increasing steadily in the park and are extending their range, principally in the North Fork section," reported the superintendent in 1937.[39] By 1942, worries about overpopulation appeared in official reports. "The elk herd in Belly River has grown to a point where any further increase will result in rapid deterioration" of elk range. Estimates of elk population, which had been as low as 337 in 1930, now ran to more than 3,300.[40]

Exactly what caused the growth of elk herds is difficult to say, but it had something to do with better-than-adequate food and cover. Winter feeding of elk, which had been practiced at Glacier for some years, can allow elk herds to increase when lack of winter range would otherwise cause some animals to starve and lower the birth rate for the herd.[41] In the various elk ranges of the park, and especially in St. Mary and Red Eagle valleys, where park officials provided tons of timothy hay every winter, the growing elk population was a positive response to administrative shaping of the ecosystem. The goal of the tourist landscape was closer to reality than ever, as thousands of elk coursed the park by the early 1940s, providing tourists with a vision of "wild" America, nature at its most vital and uncontrolled.

But the surge in elk population proved to be even less controllable than anticipated. Once the herds reached several thousand, they continued to grow, as did their demands on the limited park ranges. Suddenly, a scramble was on to find ways of reducing the number of animals. The Lewis and Clark National Forest immediately to the south of the park was a favorite hunting ground for many Montanans and tourist hunters. In 1920, the park superintendent had warned that unless the park were extended east and south, to encompass more winter range, the elk would not increase; in 1943, park officials negotiated with state game authorities to allow an extended elk hunting season in the adjacent forest with the aim of reducing park elk herds as they crossed out of the park to find winter range. Because the winter was mild, relatively few elk left the park, and the effort failed.[42]

By the mid-1940s, Glacier Park was one of several national parks that reported range damage and starvation of elk and other game.[43] In the Double Mountain area, older elk starved to death in heavy winter snows, as rangers reported with approval "nature has taken a hand in depleting the herd."[44] In a draft summary of wildlife conditions for 1944, the park superintendent reported to the chairman of a congressional committee that "Large mammals, with the exception of sheep, grizzly bear and white-tailed deer have shown a gradual increase since 1932 with elk showing the greatest increase." Some damage to the forest—like that caused by elk rubbing their antlers on trees—was unavoidable. But herds in the southern end of the park were creating more serious problems. Glacier Park was widely known for its guided pack tours, where tourists took extended trips on horseback to view the park's natural wonders. "Several sections of park trails have become impassable by pack stock due to heavy usage by elk going to and coming from the natural licks." Unfortunately, no solution had appeared yet, as the late-arriving winter of 1943 had allowed elk to remain in the park highlands until the extended hunting season in the adjacent Lewis and Clark Forest was over. The outlook for the future was not bright. "During winters of early heavy snow, the extension of the elk season may have some effect on the elk population in the park though it is doubtful."[45]

If the elk population in the southern end of the park was a problem in the early forties, complaints about the park's east side were noticeably absent. Here, mounting concerns about elk overpopulation had caused a complete reversal in the park authorities' appreciation of the "Blackfeet problem." The park superintendent wrote in 1944, "The migrating of elk upon the Blackfeet Indian Reservation and the usual take has served to keep the herds east of the continental divide within reasonable and desirable limits."[46] Once viewed as instigators of an environmental crisis, the Blackfeet and Cree hunters of the reservation had become an unintentional bulwark of ecosystem management.

But even the supposed rapacity of the Babb and St. Mary hunters could not keep elk numbers down. By 1946, rangers were reporting that "maximum use of winter range has been reached" on the east side of the park, where they counted 900 elk and estimated the total population at more than 1,100.[47] Throughout the park, there was a curious inversion of administrative perceptions about elk and the threats associated with them. Once, park officers had worried about protecting park elk, which was park "nature," when it moved beyond park boundaries. Now they became preoccupied with protecting the property of nearby residents from ravenous elk. In 1947, local ranchers to the north of Glacier complained so much about damage to haystacks and range that authorities considered fencing the park to keep the elk within its boundaries.[48]

Inevitably, however, the only effective way to control elk numbers was through hunting. After establishing sample plot exclosures and surveying the range, biologists determined that plant communities were in danger of being seriously damaged by hungry deer and elk. The Park Service arranged with the Montana Fish and Game Commission for another extended elk and deer season along the park's southern boundary. Such measures were at best only a partial solution. Elk on the North Fork of the Flathead River ranged entirely within the park, where there was no open season. As a result, the animals there were "becoming too numerous . . . and are rapidly depleting the range and driving the white-tailed deer out by depleting their food supply."[49]

By the following year, biologists were portraying the park's growing elk population as a critical problem. Noting that the total elk population for the park had been 468 in 1928, and that it had surpassed 3,000 by 1942 and was probably hovering near that now, superintendent J. W. Emmert warned that "The elk winter range is now supporting more animals than its carrying capacity." Deer were also stripping their ranges. Exclosure studies (in which sample plots were fenced to keep out elk) showed that elk had so heavily utilized parts of their winter range that in some places palatable vegetation no longer grew and had been replaced by weeds.[50]

In this context, old worries about Indian hunters continued to diminish. A park press release claimed in 1947, "Elk on the east side of the park which drift out on to the Blackfeet Indian Reservation are legitimate game for the Blackfeet Indians who are permitted to hunt all year round. Usually surplus game on the east side are kept to more or less normal numbers by the take on the reservation."[51]

Park documents still reported Indian elk hunts in which a great deal of elk were taken, but now with a tone of approval rather than impending environmental catastrophe. "Part of the St. Mary herd moved out through Babb during the winter and the Indians reduced them by about 85 head," noted the park's annual wildlife report for 1950. "With a combination of the management program being carried on and the winter loss, the elk situation in Glacier appears to be improving and approaching to normal."[52] In 1951, park officials were relieved to report: "The elk population is now down to where there is little danger of causing damage to the winter range."[53] That year there was even a momentary return to older, more familiar appraisals of Indian hunting. "The

St. Mary [elk] herd has not quite recovered from the severe winter and heavy slaughter by Indians two years ago."[54]

But the optimism was short-lived. In 1953, the park wildlife census reported that the elk count for the park as a whole had increased by more than eight hundred in two years. The winter ranges were again showing overgrazing. "The St. Mary herd . . . shows a decided increase which will require management measures."[55] Soon, park biologists were warning that the St. Mary elk herd—which now numbered almost 1,000 animals—had to be reduced by two-thirds to prevent severe damage to the range in St. Mary Valley. Other areas of the park, particularly the Middle Fork of the Flathead River, had a similar predicament. In both places, early attempts at solving the problem of elk overpopulation hinged on hunting the animals as they moved outside the park in the fall. On the east side, at St. Mary, hunters were of course those from Babb and St. Mary. Park rangers made every effort to move the elk to the park boundary by baiting them with hay and "hazing" the herds from behind, attempting to drive them out.[56] With the rangers pushing them from behind, the officials hoped that enough animals would cross the park boundary and be killed by waiting Indian hunters to reduce the herd substantially. In a remarkable reversal from only a generation before, Indian hunters were now an integral part of the official elk management plan.

After decades of warning that the Indians killed too many elk, park authorities now worried that the Indians were not killing enough. Superintendent J. W. Emmert summarized the problem in a remarkable letter to his superiors in 1953.

> As you know, the entire east boundary of the Park is bordered by the Blackfeet Indian Reservation where hunting by Indians is permitted year round. From fifty to two hundred elk are taken by Indians each year, with about fifty per year being the usual number we can expect to be killed. Occasionally, an unusually severe storm or deep snow conditions will drive a large number of elk out on the Reservation and then a good reduction is obtained. In the past two years, the increase of elk has been greater at St. Mary than any other place in the Park. Unless a heavy movement of elk from Park to Reservation occurs, we are faced with serious range damage at St. Mary in the coming year.[57]

The limited elk take by hunters at St. Mary and elsewhere around the park brought park officials to a distasteful solution: killing animals in the park. When fired upon, elk herds often disperse. The park service now hoped that by shooting a small number of elk, they could move the animals out to areas where Indian hunters could get them.[58]

Fraught with political difficulties, the "direct reduction" solution was a most unattractive option. The Park Service combined it with heavy salting of areas beyond park boundaries to draw the elk out of the park and the continuation of long hunting seasons on adjacent national forests. Sportsmen in particular were opposed to the Park Service killing animals under its protection. Park authorities were anxious not to rile these sporting interests because their cooperation with Montana officials was essential to securing long hunting seasons around the park. For this reason, the earliest direct reduction was attempted on the herd at St. Mary, "where the problem appears to warrant first trial from the standpoint of increased [elk] numbers and proximity to the Blackfeet Indian Reservation." Whereas the Blackfeet Reservation had long been problematic for the park because it was the only adjacent area where sportsmen conservationists had no influence, now it was a boon to elk reduction plans for the same reason. Recreational hunters in the national forests might complain about park rangers shooting elk to drive them across the park line. At St. Mary, "white men cannot hunt on the Reservation so there should be no complaint from sportsmen in this vicinity."[59]

During the winter of 1953–54, the park's elk reduction plan went into effect at St. Mary. As the hazing operation began, superintendent Emmert wrote a memo to his superiors warning that "The elk herd at St. Mary is now estimated at 900 head—an actual count of 600 was made last week. The range is in worse condition than ever before. All available grass has been eaten and browse plants are heavily used. If our present efforts fail, our range will suffer heavy damage this winter."[60] Aiming to reduce the St. Mary herd to 350, the animals were harassed continually from January 11 to March 18. Rangers launched flares and grenades to startle elk into crossing the park line and shot more than two dozen animals in an effort to control the movement of the herd. The herd dispersed, and "small bunches left the Park during the night and drifted onto the Reservation where the Indians made an impressive reduction."[61]

But the hazing was nearly a fiasco. Instead of moving out to winter range, many elk, alarmed at the noise of fireworks, turned around and headed for cover back up the St. Mary Valley. The park superintendent estimated that the herd was reduced to 400 animals, but that even this population "is greater than the range will support. Damage to vegetation during the late winter and early spring showed over-utilization and damage to forage crops as well as cotton-wood and conifers."[62]

Hazing continued through the mid-1950s, with park authorities worried that the elk population would surge again if they let up for even a year. In the summer of 1955, park authorities counted 270 elk in the St. Mary Valley and concluded that "the constant harassing did some good" and that it "should be an annual occurrence in order to keep this herd at a low level so as to be compatible with the range."[63] Of course, harassment by itself did not provide a solution; only the Indians killing the "surplus" elk did that. In 1963, one step of the program for "long range management" of Glacier Park's east side wildlife was to "encourage migration of elk herds to the Blackfeet Indian Reservation from specific drainages where animal overpopulation exist[s]."[64] After decades of reinforcing the invisible boundary between park and reservation, authorities now attempted to weaken it.

But the park's new impetus for elk reduction did not end the decades-long contest between Indian communities and national authorities for control over game. Indians continued to resist park authority even as rangers dutifully tried to herd elk across the park line into the sights of waiting Indian hunters. In particular, Indians remained deeply opposed to any centralized control over hunting on the reservation. This hostility was partly a response to continuing attempts by federal authorities to impose some kind of conservation program on the reservation, even at the height of the elk reduction. For park officials, it was not enough that Indians were willing to kill elk. Indeed, the goal of reducing elk numbers was strictly short-term. The long-term goal was to *control* the elk population, or in the official lexicon, to "manage" it. Elk management required controlling how much food the animals had and where they spent their winters, and collecting accurate data on as many aspects of elk life and death as possible. In short, it required a conservation program on the reservation.

Park aspirations were conditioned in part by experience with the state of Montana. In the national forests adjacent to Glacier, state game officials responded positively to park requests for extended hunting seasons. These cooperative agreements created a unified elk management program across park boundaries. Each year, the state of Montana provided data for the park regarding numbers of elk killed, and (although these numbers may not always have

been accurate), park officials could then make calculations of herd size, breeding potential, and likely reduction needs in the following year.

Indians provided no such data, and as far as park officials could tell, their
hunting was uneven. At times hunters in Babb and St. Mary took enough elk to
meet park management goals; at other times, the elk stayed within the park and
returned to it continually until the ranger at St. Mary reported in frustration,
"Indian hunters have not killed more than 10 elk all winter."[65]

Park authorities hoped that the east side elk herds could be managed much
like herds that migrated to the south. In this vision of the future, Indians would
implement their own conservation regime on the reservation, compiling records of hunting licenses sold and elk taken, creating a bureaucratic agency that
would cooperate with the Park Service in setting bag limits and season lengths
on the reservation.

Toward that end, park authorities repeatedly encouraged the Blackfeet to
implement their own conservation program. The first park efforts in this regard
were aimed at raising elk numbers. In the early 1930s, park superintendent
Scoyen appeared at tribal meetings and "tried to explain" to the Blackfeet "that
game control in this area would be of very great benefit to all of them concerned." On hearing that one man living near Babb had killed seventeen elk in
one year, he preached that such hunters were "taking sixteen of these elk from
sixteen other people"—a message that must have sounded strange in a community where hunting was more subsistence than sport and where people often
shared elk meat with the needy. Scoyen hoped that tribal authorities would
create a kind of buffer zone between the park and the reservation by closing a
six-mile-wide strip along the boundary for eleven months of the year—effectively providing the long sought-after extension of the park to the Blackfeet
Highway. His efforts seemed to pay off in 1937, when the newly organized
Tribal Business Council promulgated the first tribal game code, establishing a
closed season on elk and limiting Blackfeet hunters to one elk per year.

These regulations were ill-enforced, however, and by the early 1940s federal observers despaired of the tribe's conservation program, which seemed
designed more to resist state incursion than to protect animals. The Indians
"buffeted game management about with reckless abandon," repeatedly refusing
cooperative agreements with state conservationists and insisting on their right
to sell their own hunting licenses to white hunters.[66] In essence, the Blackfeet
game code, as it developed after the 1930s, was designed in part to bound a

tribal hunting commons—from which white hunters were usually excluded—
and as a bulwark against outside meddling by the Park Service or state game
officers.

Park Service longing for a "real" conservation program on the reservation
did not diminish even as elk populations surged. In 1954, at the height of the
elk reduction program in St. Mary, park officials were still hoping for "regulation of shooting seasons by the Blackfeet Indians on their reservation" as part of
the greater elk management program for the park's east side.[67] Even with the
increasing elk populations sending more animals than ever onto the reservation, Indians refused. One frustrated federal conservationist, surveying Blackfeet opinion on wildlife conservation in the early 1940s, estimated that "the
principle of wildlife conservation is opposed by about 95 per cent of the Indian
population."[68] In fact, by the 1960s, the Blackfeet had eliminated the long-
unenforced tribal bag limit on elk, asserting a local Indian right to determine
Indian hunting behavior.

Troubles with the Indians aside, by the 1950s park goals for managing the
St. Mary herd were the reverse of park goals in the 1920s. "It is hoped that,
eventually, these elk will develop a migratory habit and leave the St. Mary valley
when winter snow conditions and constant harassing are combined to make
foraging difficult." The newest obstacle to this plan was a perceived reluctance
among the elk to leave the park. Officials took the animals' wariness to be a
consequence of experience: elk knew that the further they moved from the
mountains, the more likely was hunting to be a threat. Thus, the old hope for a
reservation buffer zone between the mountains and Indian hunters remained
very much alive, with a dramatically different goal. Where authorities once
hoped to keep hunters away with a buffer zone, now they wanted it to encourage elk migrations to the reservation.[69]

Indian resistance to park authority was such a dominant characteristic of
local life that it came to characterize the oral history of these years as well. In local lore, there is no memory of rangers trying to chase elk out of the park. Commonly, locals recall park rangers using fireworks and shooting at the animals to
chase them back in.[70] In part, this popular recollection probably stems from
eyewitness accounts of the elk hazing program. When the rangers launched
rockets at the elk herd, many animals turned and ran back up the St. Mary
Valley. No matter what rangers and other park officials might have said they
were doing, locals knew from experience that the park authorities would deny
them game at every opportunity. When they saw rockets exploding and elk

running deeper into the park, they interpreted it as a new twist on an old effort to keep elk off the reservation.

Resistance to conservation therefore remained a defining feature of life at Babb and St. Mary. Being local required a large degree of dependence on the national park, but it also meant defying park managers. This was true in other communities around the park as well, but at Babb and St. Mary it was strengthened by Indian experience, especially Blackfeet experience, of federal power. In the eyes of the Indians, the 1896 treaty gave them the right to hunt in the park's east side, and references to "conservation," "game management," and "wildlife protection" were merely euphemisms for breaking that promise. The hunting rights provisions of the 1896 agreement remained a flashpoint for park-Indian conflicts throughout the twentieth century, and poaching in the park was only one manifestation of local anger. For generations, residents of Babb and St. Mary contested the location of the park boundary, maintaining that some land claimed by the park belonged to the Indians outright.[71] The mushrooming elk populations in the St. Mary Valley at mid-century could change park perspectives on reservation hunting, but they could not subdue Indian challenges to park legitimacy in the ceded strip.

The contest between local community and national authority had long revealed underlying differences in the ways people understood elk. To the park authorities they were a tourist attraction, and part of the park's "natural" landscape. To the people of Babb and St. Mary, they were food, hides, and clothing, and could be exchanged for cash; moreover, hunting the animals expressed ancestral claims to the land. The contest over these animals was always sharpened by their tendency to wander from the park, where those in power understood elk in one way, to the Indian reservation, where the powerful, training their rifles on the animals, understood them in another. The boundary between park and reservation was meant to separate two commons regimes. But in fact, the elk linked the separate systems of park and reservation, signifying that elk were not tourist attraction or meat, but both.

The irruption of elk at St. Mary created a strange new world on the western edge of the reservation. After decades of trying to keep elk in the park and away from Indians, rangers tried to chase them out of the park and to the Indians. Indians, witness to the spectacle and still suspicious, took large numbers of elk and refused to abide by park service recommendations to "regulate" the hunt. It was a grudging arrangement, but it seemed somewhat effective for controlling elk herds.

The dramatic rise in elk numbers, reflecting the changing face of the contested commons, in this sense created a new setting for old disputes. In January 1956, district ranger Barnum reported to park headquarters that 150 elk had left the park and that Indian hunters had killed 59 of them. The elk had "scattered out on the rolling hills east of the park, a small bunch going out towards Duck Lake," to the east. Other, small bands of elk left the park in subsequent weeks. "There is doubt that a few of the elk which left the park may return," wrote the district ranger, but there was still room for optimism. Because "the Indians have accounted for 77 head. . . . It is quite certain that the small herd which went out toward Duck Lake will never return."[72]

With the small herd went older, simplistic notions of controlling elk populations. For decades, park authorities warned all who would listen that elk were on the verge of extinction at Glacier; for decades, they sought every advantage in their campaign to separate elk from Indians. Legal decisions and ranger deployments never seemed to constitute the kind of wall the Park Service hoped for along the park's east side. After all the directives and anguish, this was a strange memorial to their efforts: rangers chasing herds to waiting Indians, rockets flaring over the backs of the retreating animals, each arcing trail of smoke perhaps to resemble nothing so much as a bridge, connecting park and reservation.

When Napi failed to kill the last elk in the world, he released a life-giving power into the world, and it was not up to people to understand or control it. Before the coming of the Americans, Blackfeet intimacy with animals and their utter reliance on them was manifest across their huge expanse of hunting grounds; after the American conquest, the utter devastation of those same hunting grounds, from the mountains to the once-teeming Sweet Grass Hills, was a fundamental cause of their own desperation. Well into the twentieth century, the presence of animals empowered the Blackfeet. Without elk to worry over, the Park Service might have paid far less attention to Blackfeet claims on the park. In a sense, the elk brought not only food and hides each fall, but also the attention of federal authorities who were often all too willing to ignore the Indians otherwise. If the result was a constant battle over hunting rights, at least the Blackfeet were still fighting for what was theirs. In a real sense, the ability of the elk to grow so numerous (and become so ravenous) meant that Blackfeet had an otherwise unthinkable leverage in their relations with the Park Service and the federal government as a whole.

If animals could empower people by their arrival, then they could be a harbinger of better times ahead. And for Indians and whites alike there was a very good sign indeed when the fall migration of 1954 was over. By this time, the herd at St. Mary was much reduced, owing to the large number of elk which had vanished into the reservation, where rangers assumed they fell before Indian guns. But at least one herd survived a migration across the entire length of the Blackfeet Reservation. They must have moved at night, and escaped hunters by hiding in river valleys far out on the reservation where no one would be looking for them. They reappeared on the other side of the reservation boundary, to the east. At Cut Bank, fifty miles from the park, locals reported a stunning event. For the first time in over half a century, elk had returned to the Sweet Grass Hills.[73]

Epilogue
Localism, Nationalism, and Nature

Blackfeet were different in many ways from other locals who confronted the rise of conservation, but in important respects their story remains suggestive of hunting struggles throughout the century. As in many other places, the changing ecology changed resource politics—but not completely. Indians continued to demand treaty rights to the ceded strip, and in the militant days of the 1960s and 1970s, young activists again challenged the Park Service in court. The presiding judge doubted the legality of the government's expropriation of hunting rights and awarded the Indians the right to enter the park without paying the entrance fee.[1] The victory was both partial and unprecedented. But oddly, rather than the expected string of follow-up claims and appeals, Indians and the Park Service have steered clear of the courts since then. Perhaps Indians and authorities alike are reluctant to submit their hard-won gains to the courts again; perhaps each side senses it has too much to lose. There are growing demands from within the Blackfeet tribe for rigorous game conservation on the reservation, and some on both sides hope for a joint management plan for the park's eastern side.

In any case, for national authorities the Glacier experience was unsettling. Social conflicts made game protection difficult, and ecological fluctuations made it more so. It became harder for authorities to pinpoint the enemies of abundance when the land kept changing beneath their feet. And, no matter how loudly or often officials spoke to the contrary, disputes between locals and central authorities were no more settled than the land. In the end, it was so difficult to separate elk from local people, or to understand them apart from local ecosystems.

For the Blackfeet, too, there were lessons in this experience. As they knew all too well, they were the most local of people, and yet they relied to no small degree on an extra-local, national benefactor, the federal government. Much as they resented the national park at Glacier, they had integrated it into their lives. Had there been heavy mining and lumbering of the region and no federal protection for the park, there would probably have been fewer elk, if any. More to the point, it was to federal courts that Indians made their legal appeals, and arguably because of their status as a federally protected people that they could not easily be dispossessed of the lands along the park boundary. As local people under national protection, the Blackfeet embodied the mixing of local and national identities that in so many ways has come to typify Indian Americans.

In other ways, this dualistic, seemingly contradictory admixture of local and national identities became characteristic of the West as a region. Its paradoxes are particularly striking in western New Mexico. The same towns that became almost hysterical at the rumor of Indian hunters a century ago are seemingly just as worried—practically paranoid—about the federal government today. In the town of Reserve, residents recently planned to burn a United Nations flag, and Forest Service agents are reluctant to step outside their offices for fear of public hostility. No town would seem more anxious to separate local living from national presence—and yet that goal remains unattainable. Many of the same residents who resent "outside" authority paradoxically rely upon it to ensure local order. Ranchers demand and receive grazing subsidies on the public lands. Lagunas and Navajos still hunt the Mogollons, and isolated ranchers still worry about Indian hunters who stalk their ranges in the autumn, unannounced and uninvited. At least as worrisome, and possibly more so, is the seasonal influx of non-Indian, tourist hunters, who are renowned for drinking to excess, leaving gates open, and shooting stock.[2] Social controls may be effective against locals who are connected to landowner communities through networks of kin and acquaintance, but controlling Indians and tourists, both of whom belong to social networks outside local society, requires the help of "outside" authorities. For all the public resentment of "federal domination," local life has become so entwined with a large federal presence that it is difficult to imagine a "non-federal" West.

The purpose of this book has been in part to explore how the growing national presence manifested itself in the hunting grounds, and to understand who gained, and who lost, when it did. The "national commons" in wildlife represents one of the most startling examples of how the nation acquired power over the most intimate of relationships between people and nature. From an early date, the West was the more federalized place in this respect, but the rest of the country soon followed. During the Depression, eastern states, hard pressed for cash, looked to the federal government to take over large tracts of vacant land— much of it superb wildlife habitat—under the authority of federal emergency programs. Although most of its landholdings were still in the West, by the end of the 1930s the Forest Service regulated forests and wildlife across the country.[3]

As in the West, individual states simultaneously cultivated and bemoaned the federal presence, and Pennsylvania was no exception. In 1940, the director of Pennsylvania's Game Commission complained bitterly to Congress about the degree of federal involvement in his state's game management efforts.[4] There was a certain poetic irony to these anti-federal sentiments in Pennsylvania. For at least one prominent conservationist, the state's wildlife had signified a special kind of connection to the West. Joseph Kalbfus, the director of the Pennsylvania Game Commission during the Seely Houk murder investigation of 1906 and for many years afterward, was born and raised an easterner. Too young to fight the Civil War, Kalbfus had gone West in the 1870s at the age of nineteen, hoping to fight Indians. Although he never did, he was entranced by the vast herds of game in his adopted home of Estes Park, Colorado. He became an avid hunter, known to his rustic neighbors as Antelope Joe. Returning to the East for good in his early twenties, his western years became the formative experience of his life, to such an extent that in his autobiography he memorialized himself in a series of frontier tropes. Early conservation of Pennsylvania wildlife became, in his recollection, an effort to make the state's hunting grounds as abundant as the West of memory. His conflicts with Italians seemed a kind of substitute for the Indian war he never fought; the receipt of Black Hand death threats in the mail, he reminisced, "stirred me to a fighting pitch I had not known since I left the Plains."[5]

His dreams of game abundance seemed more than fulfilled by the mid-1910s, as hungry deer spilled out of Pennsylvania forests to overrun farm fields. Having hoped to recapitulate a part of the "unsettled" West in the forests and fields of the East, he saw too late the massive unsettling this development would bring to his conservationist coalition. We can only wonder what he might have thought as federal managers became more and more responsible for Pennsylvania's wildlife in the 1930s, making the state seem "western" in ways he never intended.

State resistance to federal conservation efforts had limited effect in the long term. The U.S. Department of Agriculture eventually rescinded some of its most controversial directives, but the national commons in wildlife continued to expand.[6] Federal powers over game grew with the Fish and Wildlife Act of 1956, the Endangered Species Acts of 1966, 1969, and 1973, the National Environmental Policy Act of 1969, and others, as well as the growth of the National Park system.[7] By the latter half of the twentieth century, federal involvement in disputes over American wildlife was commonplace, even when the animals in question were on private land, even in the East.

Where the federal government took a strong hand in governing the West, the region set the pattern for the rest of the country. The embedding of the federal government not just in wildlife issues but in daily life has become, to some degree, typical of localities everywhere in America. Whether or not this development is healthy for communities, it is striking how much the national government has become part of our lives, and how much our sense of ourselves and our "local" communities has come to include a strong federal presence—a post office with overnight delivery, guaranteed student loans for the community college, a defense research contract for the state university—in many cases without our even realizing it.

Constable Manning could not have known it, but in a curious way his patrol that morning in July 1895, when he hoped to bring in the federal government to separate Ben Senowin and the other Bannock Indians from elk, was in a sense a precursor, a small bridge among many similar local actions that ultimately led to a much larger federal presence in wildlife matters. In this sense, the posse was a curious forerunner to greater conservationist endeavors, to better-intentioned laws like the Endangered Species Act. In this connection conflicts like those at Fall River—though remote in time and geography— become essential to an understanding of twentieth-century America. As Samuel P. Hays has observed, conflicts between centralization and decentralization, between expertise and localism, are not peculiar to conservation but typical of modern American society.[8] Seen in the context of the eventual explosive growth of national power and presence in local life, America's bitter struggles over wildlife conservation become hallmarks of the modern experience. Ranchers and hunters who watched carefully for federal agents on the slopes of the Mogollons or in Glacier's deep valleys had this much in common with other Americans: the federal government was emerging as a major, reorganizing force in their lives.

Locals could respond in many ways to this development, but critical in their decision was the issue of abundance, the ability of the local commons or extra-local management regimes to provide enough of what local people define as essential for "the good life." Sometimes, conservationists provided well, as when the national forests allowed grazing for cattle ranchers in the vicinity. But for nomadic sheep ranchers, shut out of the national forests in deference to cattle owners, the national commons severely curtailed the customary grazing freedoms inherent to local notions of abundance. To other rural people—the Hispanos whose communal lands were alienated to the Forest Service in northern New Mexico and Indians whose hunting and fishing areas fell under control of remote government officers—the loss of the local commons could be devastating.[9] Hunting in national parks and other violations of game laws were only some of the myriad ways that dispossessed locals resisted the onset of state authority over the local commons. In diverting water from state and federal irrigation projects, cutting timber and grazing cattle without permits, and refusing to pay for customary use rights on federal lands, locals made resistance to conservation a way of life.[10]

There were many who came to believe that the presence of national authorities diminished the autonomy of local communities, and for these people the past held a particularly bitter lesson: the federal grip on the West came about partly as a result of the repeated discounting and abandonment of localism by local people themselves. It was not just that elites among them called for federal assistance. More pervasively, Americans expressed a willingness to temporarily de-localize themselves, to relocate, and thereby created many of the problems state and federal governments were expected to solve during the late nineteenth century. As millions of foreign nationals and native-born Americans migrated westward from Europe and the East Coast, they temporarily cast off the constraints of home and local identity. Taking up residence elsewhere, those who came to the hinterlands inevitably challenged prior notions of proper relationships to game animals. New and old residents were frequently divided by language, culture, and divergent notions of land and the good life. To new locals, whether Italian immigrants in Pennsylvania or Anglo ranchers in New Mexico, the prior locals were often a hindrance or worse, interfering with the orderly use of the land; to older locals, the farmers of Lawrence County, Pennsylvania, and the Indians of New Mexico and Montana, immigrants were the problem, callously upsetting local traditions of land use. For each camp, the wildlife commons seemed unsettled by the activities of the other.

At such times of tense confrontation, elites found localism wanting. Pennsylvania farmers called on the distant Game Commission, New Mexico ranchers turned to Santa Fe authorities and the federal government, and businessmen in northwestern Montana allied themselves with bureaucrats in Washington. This was a singular paradox: the creation or restoration of a particular local lifeway so often seemed to require the help of extra-local agencies.

If conservation was a way of reordering the land to suit elite local interests, it is striking how often those interests involved courting extra-local visitors. Emerging economies of tourism became a powerful force that allowed local elites and conservationists to utilize wildlife in new and important ways. Throughout the West, federal authority was repeatedly placed at the disposal of those who built tourist recreational paradises. Partly through the initiatives of local elites, local land and game in southwest New Mexico, northwest Montana, and elsewhere became to a significant degree the property—or the privilege—of a national, middle- and upper-class tourist public.

In an essential precursor to this appropriation of local game, dominant interests across the country installed legal codes to control hunting by minority groups. But between East and West, different conditions led to divergent strategies, and the ramifications were profound. Insofar as degrees of "localness" were concerned, eastern and western elites were remarkably distinct. Immigration caused unsettling social changes across the continent, but in the West, it was immigrants, not prior locals, who came to wield state power over the hinterlands, and they did so by appealing to federal authorities more than their eastern counterparts did.

In part, western immigrants' support for conservation was a way of addressing their own self-consciousness as newcomers. Suffocating prior local claims beneath the blanket of extra-local, federal authority became a way for new immigrants to assert control over mountains, forests, and game, and the Indians and Hispanos who had ancient claims to them. With federal rangers patrolling the forests and federal agencies to complain to, the immigrants felt marginally more comfortable in this vast country. They felt more at home—more local.

Partly because of this pervasive facility at using the national government to serve local interests, the West is still the most federal of regions. Here, three-quarters of a billion acres (roughly the size of India) belongs to the federal government.[11] And it is once again the site of a large-scale reordering of local lifeways. Millions of computer programmers, artists, and professionals have emigrated to the intermountain region, and longtime residents find their assumptions about the region's vast public holdings challenged in new ways. Older economies of resource extraction wane, new economies of information grow, and, across the region, changing patterns of settlement and market demand are radically altering local life.

Perhaps the most visible change is in the public attitudes toward western lands, which are more than ever seen as recreational landscapes. At Jackson Hole, Wyoming, nearly a century after Constable Manning and his posse set out to protect elk for tourists, another rancher, Rod Lucas, set out to move his cattle to their customary summer range. A stream of tourist traffic forced his herd off the road, and Lucas gave up in disgust. "The cow business as everybody knew it is gone. I need to get out of this tourist country." Silver City, New Mexico, is now home to the Artist-in-Residence Espresso Bar and Gallery; where miners once toiled for mineral and metal, mountain bikers and hikers demand meadows and trails. Older residents resist, often through federal agencies that have long served their interests, especially the Forest Service and the Bureau of Land Management. Ranchers demand payment for grass consumed by public elk on "private" leaseholds, and some want a final sell-off of the public lands, ostensibly in keeping with American traditions. Whatever their differences, all agree that the future of local life depends to a large degree on federal action. The national commons is being reordered, again.[12]

The Blackfeet were disenchanted with tourism in their old hunting area long before Rod Lucas was, a fact that suggests how similar present-day confrontations are to older ones. Often, old-timers consider new immigrants to be "extra-local," and some—especially fugitive Californians—are considered even stranger and less assimilable than others. Implicit in much "old local" criticism is the notion that "latte sippers" and computer consultants cannot possibly be "real" locals, for the West was not settled by "people like them." Whatever the merits of the argument, it seems undeniable that when it comes to land use, very little about the West was ever "settled."

In no small measure, debates about the future of the West hinge on the degree of federal influence on local life. Antigovernment agitators—some local, some not—have apparently settled on localism as the proper response to what they see as an overbearing, insensitive federal government. The so-called Counties Movement seeks to invalidate all federal regulations over land use by claiming the public lands as local property. Deals with cash-starved locals have often allowed large corporations cheap access to resources, and for this reason the corporate-sponsored Wise Use Movement cynically demands "local prerogative" over public lands. The most radical proposals come from fringe elements of the far right, the self-styled militias, who have seized upon western symbols like the six-gun and the stetson to mount a violent campaign against what they consider federal tyranny.[13]

Economic and other flaws aside, for our purposes the most striking feature of this insistent localism is how much it defies historical tradition. Whatever the failings of the commons, privatization of the public domain would effectively hand over to local elites (and many corporations) an enormous resource long held in common by communities, local and national. And for all their anxiety over preserving "local custom and culture," ranchers, in demanding a "Fed-free" range, violate local tradition. As we have seen, federal authority arrived in the West with the same local communities who now claim an exclusive right to the land, and often at their request. As much as ranching itself, federal regulation is a local custom, embedded in local culture, however it may be defined.

Many of the most fervent dissidents are recent immigrants to the region who eschew even customary local police authority. Such extremism often falls afoul of locals' sense of propriety. Ironically, such conflicts lead only to less local autonomy. When the FBI arrived in Jordan, Montana, in 1996 to arrest the Freemen, it was at the behest of locals who were fed up with Freemen threats, fraudulent checks, and extremist localism.[14]

Of course, localism is not just a western emotion; it is an American tradition, having been a part of our political debates at least since Nathaniel Bacon faced off against Governor Berkeley in colonial Virginia. But the sentiment finds particular resonance in the West, if for no other reason than it is the most palpably "de-localized" of America's regions, seemingly the most alienated from local control. Partly this comes from a heritage of heavy extra-local demand. For the past century, thirsty, growing metropoles have soaked up more and more of the region's scarce water, and the U.S. armed forces have appropriated enormous stretches of territory as training and testing grounds. Private capital is no less "de-localizing": the immense power of multinational corporations has long lent the West an air of colonialism. But equally telling are issues of ownership and control. Monument Valley, the Grand Canyon, and the Mojave Desert are vistas of mythic importance in American identity, cultural symbols in their own right that Americans in general have no trouble calling theirs. But among those who live nearby, there is a decided ambivalence about these sites. They are local lands, part of home; but they usually belong to the federal government, to the public, and are beyond the control of local people. So, too, in less appreciated, less popular rural places. For these reasons, westerners are as likely to have a working relationship with federal agents as with state or even local leaders. And the word *local* exerts a stronger-than-usual appeal. In addition to its many other meanings, *local* implies a degree of comfort in the landscape, a sense of being at home in a place others find strange. Depending on the context, it can convey a degree of entitlement or even ownership. Small wonder that localism is especially strong in this region, where everyday living depends so much on exchanges and contracts with the least local authorities.

Much as they helped to create this de-localized West, conservationists did not see themselves as enemies of local communities. Expropriating power from them was only a means to an end. Their goal—even if they failed to achieve it—was to create abundance for locals everywhere, through scientific management of the land, and the market's talent for rational distribution. In turning to the federal government to accomplish this end, conservationists combined old political traditions—"the general welfare" had long been an area of federal interest, after all—with new innovations and developments. For many at the turn of the century, the Civil War had established the federal government as the preeminent agent of reform and liberation. Compared to the emancipatory power of national authority, local government seemed backward, provincial, and corrupt.[15]

Heirs to this legacy, conservationists wedded it to evolving notions of American nature and culture. During the nineteenth century, landscape artists of the Hudson River School had radically reshaped the popular understanding of natural landscapes. Americans, long self-conscious about the absence of American cultural monuments to rival those of Europe, found a new self-confidence in the paintings of Thomas Cole, Frederic Church, and others. Here, America's abundant, pristine landscapes became manifestation of God's word, proof of the divine origins and destiny of "Nature's Nation," the United States of America.[16] Once the Civil War had validated the bonds of federal power as cleansing and liberating, the notion of using it to protect and preserve America's abundant nature seemed fitting and necessary.

Thus, to Gifford Pinchot and Theodore Roosevelt, as well as to Page Otero, Joseph Kalbfus, and maybe even Constable Manning, conservation was the vanguard of national freedoms, serving purposes at once material and holy. A vigorous state could parcel out God's nature, preserving the heritage of the frontier and guaranteeing a level of material wealth high enough to preclude corporate monopolies, inequity, and class strife. Proponents might say that conservation would allow the nation to achieve its dual purpose, the propagation of democracy and Christianity.[17]

In practice, as seen in the situations discussed in this book, what abundance conservationists achieved came at a cost. Whether they were making

Pennsylvania's hinterlands safe for farmers and sport hunters, creating a "free" New Mexico hunting ground for middle-class hunters, or opening a national park in Montana for the enjoyment of tourists, conservation secured resources for many at the expense of the few. Its champions even reveled in this, deriding their opponents as rich, pampered corporations, greedy "special interests" seeking to deny the public its birthright.

Many opponents of conservation were wealthy, but labeling all their critics as plutocrats was also a way for conservationists to deflect attention from their program's tendency to deny freedoms long inherent to local life. Among the people who suffered most were the rural poor. There was no escaping this contradiction: where the "national commons" secured liberties for some, it also brought a significant degree of coercion. At times, this tendency inspired vigorous local opposition, for many people agreed with the New Mexico old-timer who said that God, not conservationists, made deer, and not for tourists but "to be used for food by the poor folks."[18]

Despite the tendency of historians since Frederick Jackson Turner's time to look for a "closure" to the frontier, examining the American West from the vantage point of the commons suggests useful connections between the nineteenth-century West and the West of our own day. Turner's frontier was, among other things, a patchwork of local commons. During the latter nineteenth and early twentieth centuries, many local commons regimes were gathered into a national pool of resources in which the national public was to retain a version of community use rights. In the process, a radical reorganizing of social relations took place in the countryside, reinforcing the power of local elites and challenging many closely held ideas about the nation and its nature.

The national commons Americans have inherited contains so many different resources under so many different management regimes as to defy generalization. But even as they fend off the most recent surge of localist sentiment, environmentalists and other supporters of federal resource management concede that conservation achieved only limited success in the area of community use rights. In many ways the greatest failing of conservationists was that the putative owners of the commons—the public—rarely and only indirectly managed it. The national parks are open to the public at large and clearly managed for, but not by, a wide class of tourist users. Today, critics charge that national forests and public lands are public only in name, that they have fallen to timber, beef, and mining corporations who have co-opted government agencies charged with protecting the resources.[19]

And critics have been quick to seize on failures of abundance. Managerial control over land was as fleeting a vision as perpetual plenty. Nowhere was this clearer than in wildlife management, where shifts in ecological dynamics sent wildlife populations spiraling up and crashing down, leaving managers to puzzle over the debris. Even bountiful hunting grounds did not necessarily satisfy everyone. For farmers and ranchers who wanted better regulation of the hunt to control "outside" hunters, a plentiful deer herd could mean more armed strangers in the countryside, more lost crops and cattle. Where the national commons brought both freedom and coercion, it also delivered abundance and scarcity, often at the same time, and often unforeseen.

Perhaps the unpredictability of wild animals was the most enduring and positive of their many attributes. Although opponents in these struggles rarely understood how or why wild animals appeared and disappeared from the land, the animals' ongoing tendency to do so forced a constant reckoning and re-reckoning of the land and the living it provided. Ultimately, partisans of both localism and nationalism sought to embed their respective causes in nature, which fortunately for both sides proved perpetually resilient. Struggles over game would not have happened without game to fight over; the ability of wild animal populations to rebound, far outstripping even the most optimistic projections of conservationists, meant that these contests could expand into new places in the years to come. As they did, they divided communities, but they also provided a reason to rethink the way people lived on the land and alongside one another, native-born and immigrant, farmer and sportsman, rancher and bureaucrat, Indian and white.

The question of how the local commons became the national commons lends a compelling quality to many stories of the West, where we see so clearly the emergence of the national from what was local. It is this dimension of Ben Senowin's story that speaks to every American, Indian and non-Indian. His is a story of murder and attempted massacre, but it is more. It encapsulates the emergence of local into state and national, of "us" into something bigger, sometimes more powerful—but often terrifying. There remain many dark and troubling questions about this transformation, not least how the nation should best administer public resources like game, wood, and water. Whether a nation can act as a community, and whether in this regard a true national commons is even possible, remain to be seen.

Notes

The following abbreviations are used in the notes.

ARBGC	*Annual Report of the Board of Game Commissioners*, Pennsylvania
BIA	Records of the Bureau of Indian Affairs
BRBGC	*Biennial Report of the Board of Game Commissioners*, Pennsylvania
GNP	Glacier National Park
NMDGF	Records of the New Mexico Department of Game and Fish
RLA	Ruhle Library Archives, Glacier National Park
SAR	*Superintendent's Annual Report*
TANM	Territorial Archives of New Mexico

Prologue

Note to epigraphs: Hamilton A. Tyler, *Pueblo Animals and Myths* (Norman: University of Oklahoma Press, 1975), 63; Laura Ingalls Wilder, *Little House on the Prairie* (1935; repr. New York: Scholastic Book Services, 1975), 2.

1. This account is taken from U.S. Commissioner of Indian Affairs (hereafter USCIA), *Annual Report of the Commissioner of Indian Affairs 1895* (Washington, D.C.: Government Printing Office, 1896), 60–80.

2. Small bands of Bannock had hunted the area each summer for much of the nineteenth century. There were no fixed hunting areas, and among the Bannock bands there were no proprietary rights to particular hunting grounds. Robert F. Murphy and Yolanda Murphy, "Shoshone-Bannock Subsistence and Society," *University of California Anthropological Records* 16 (1961): 329. See also E. Adamson Hoebel, "Bands and Distributions of the Eastern Shoshone," *American Anthropologist* 40 (1938): 410–413.

3. Charles Elton, *Animal Ecology* (New York: Macmillan, 1936), 34.

4. USCIA, *Annual Report of the Commissioner of Indian Affairs 1896*, 60–66. Quotations from p. 62. The Supreme Court decision was handed down in *Ward v. Race Horse,* 163 U.S. 504.

5. Frederick Jackson Turner, *The Frontier in American History* (New York: Henry Holt, 1920), 1.

6. Garrett Hardin, "The Tragedy of the Commons," *Science* 162 (Dec. 13, 1968): 1244.

7. Ibid., 1244, 1247–1248.

8. For an introduction to this literature, see Garrett Hardin and John Baden, eds., *Managing the Commons* (San Francisco: W. H. Freeman, 1977); and Bonnie J. McCay and James Acheson, eds., *The Question of the Commons: The Culture and Ecology of Communal Resources* (Tucson: Arizona Studies in Human Ecology, 1987).

9. Terry L. Anderson and P. J. Hill, "From Free Grass to Fences: Transforming the Commons of the American West," in Hardin and Baden, eds., *Managing the Commons*, 200–216; Arthur F. McEvoy, *The Fisherman's Problem: Ecology and the Law in the California Fisheries, 1850–1980* (New York: Cambridge University Press, 1986).

10. Resource policy theorists have used Hardin's model in an attempt to explain certain aspects of the economic history of the American West as a "frontier." Their conclusions tend to be supportive of Hardin's thesis. My own are much less so, as will become apparent in the paragraphs that follow. Terry Anderson and Peter Hill, "From Free Grass to Fences"; also, by Anderson and Hill: "Property Rights as a Common Pool Resource," in John Baden and Richard L. Stroup, eds., *Bureaucracy vs. Environment: The Environmental Costs of Bureaucratic Governance* (Ann Arbor: University of Michigan Press, 1981), 22–45; "The Evolution of Property Rights: A Study of the American West," *Journal of Law and Economics* 18 (1975): 163–180. Also, R. Taylor Dennen, "Cattlemen's Associations and Property Rights in Land in the American West," *Explorations in Economic History* 13 (1976): 423–436.

11. Turner, *Frontier in American History,* 11.

12. My criticism of Hardin's thesis comes largely from Bonnie J. McCay and James M. Acheson, "Human Ecology of the Commons," in McCay and Acheson, eds., *Question of the Commons,* 1–36, and by other authors in the same volume. Criticism of the Turner thesis is now so pervasive as to be characteristic of American history as a field. For an economic analysis of the hidden costs of "free land," see Clarence H. Danhof, "Farm-Making Costs and the 'Safety Valve': 1850–1860," in Vernon Carstensen, ed., *The Public Lands: Studies in the History of the Public Domain* (Madison: University of Wisconsin Press, 1963), 253–296.

13. David Potter, *People of Plenty* (Chicago: University of Chicago Press, 1954).

14. Julian Steward, "Basin-Plateau Aboriginal Sociopolitical Groups," *Bureau of American Ethnology Bulletin* 120 (Washington, D.C.: Smithsonian Institution, 1938), 33, 200–201.

15. McEvoy, *Fisherman's Problem.*

16. My definition of local commons is offered below. But see David Feeny, Fikret Berkes, Bonnie J. McCay, and James M. Acheson, "The Tragedy of the Commons: Twenty Two Years Later," *Human Ecology* 18 (1990): 4–5; McCay and Acheson, "Human Ecology of the Commons," 16, 27; also Devon Pena and Eric Boyd Del Balso, "Paradigm of the Homeland Commons: Local Ecosystem Stewardship in the Upper Rio Grande," unpublished manuscript.

17. S. V. Ciriacy-Wantrup and R. C. Bishop, " 'Common Property' As a Concept in Natural Resources Policy," *Natural Resources Journal* 15 (1975): 713–727.

18. Steward, "Basin-Plateau Aboriginal Sociopolitical Groups," 200–201.

19. The issue of "territoriality" in defining commons regimes has been a major focus of various anthropologists. See James M. Acheson, "The Lobster Fiefs Revisited: Economic and Ecological Effects of Territoriality in the Maine Lobster Industry," in McCay and Acheson, eds., *Question of the Commons,* 37–65; also, in the same volume, Fikret Berkes, "Common Property Resource Management and Cree Indian Fisheries in Subarctic Canada," 66–91, esp. 71.

20. USCIA, *Annual Report of the Commissioner of Indian Affairs 1895,* 75.

21. Again, I owe an intellectual debt to Arthur McEvoy's work on this point.

22. Dan L. Flores, "Bison Ecology and Bison Diplomacy: The Southern Plains from 1800 to 1850," *Journal of American History* 78 (1991): 465–485.

23. Stephen J. Pyne, *Fire in America: A Cultural History of Wildland and Rural Fire* (Princeton, N.J.: Princeton University Press, 1982), 71–83, 252–253. See also Richard White, *The Roots of Dependency: Subsistence, Environment, and Social Change Among the*

Choctaws, Pawnees, and Navajos (Lincoln: University of Nebraska Press, 1983), 316–317.

24. Richard E. McCabe, "Elk and Indians: Historical Values and Perspectives," in Jack Ward Thomas and Dale E. Toweill, eds., *Elk of North America: Ecology and Management* (Harrisburg, Pa.: Stackpole Books, 1982), 65–68.

25. William Cronon, *Changes in the Land: Indians, Colonists, and the Ecology of New England* (New York: Hill and Wang, 1983), 13.

26. Samuel P. Hays, *Conservation and the Gospel of Efficiency: The Progressive Conservation Movement, 1890–1920* (Cambridge, Mass.: Harvard University Press, 1959); Robert Wiebe, *The Search for Order, 1877–1920* (New York: Hill and Wang, 1967); Gabriel Kolko, *Railroads and Regulation, 1877–1916* (Princeton, N.J.: Princeton University Press, 1965).

27. William Cronon, *Nature's Metropolis: Chicago and the Great West* (New York: W. W. Norton, 1991), 46–54.

28. Richard White, *"It's Your Misfortune and None of My Own": A History of the American West* (Norman: University of Oklahoma Press, 1991). Also Michael J. Bean, *The Evolution of National Wildlife Law* (New York: Praeger, 1983); Hays, *Conservation and the Gospel of Efficiency;* also by Hays, *Beauty, Health, and Permanence: Environmental Politics in the United States, 1955–1985* (New York: Cambridge University Press, 1987), 13–22.

29. Gifford Pinchot, *The Fight for Conservation* (1910; repr. Seattle: University of Washington Press, 1967), 103.

30. William deBuys, *Enchantment and Exploitation: The Life and Hard Times of a New Mexico Mountain Range* (Albuquerque: University of New Mexico Press, 1985).

31. E. P. Thompson, *Whigs and Hunters: The Origin of the Black Act* (London: Allen Lane, 1975); Douglas Hay, "Poaching and the Game Laws on Cannock Chase," in Douglas Hay et al., eds., *Albion's Fatal Tree: Crime and Society in Eighteenth-Century England* (New York: Pantheon Books, 1975); P. B. Munsche, *Gentlemen and Poachers: The English Game Laws, 1671–1831* (New York: Cambridge University Press, 1981); Harry Hopkins, *The Long Affray: The Poaching Wars, 1760–1914* (London: Secker & Warburg, 1985). Colonial and African historians have also utilized game law and conservation history in innovative ways: John Mackenzie, *The Empire of Nature: Hunting, Conservation, and British Imperialism* (New York: Manchester University Press, 1988); *Journal of Southern African Studies* 15, no. 2 (January 1989), "Special Issue: The Politics of Conservation in Southern Africa." For anthropological treatment of hunting issues in Africa, see Stuart Marks, *Large Mammals and a Brave People: Subsistence Hunters in Zambia* (Seattle: University of Washington Press, 1976), and, also by Marks, *The Imperial Lion: Human Dimensions of Wildlife Management in Central Africa* (Boulder, Colo.: Westview Press, 1984).

32. James C. Scott, *Weapons of the Weak* (New Haven: Yale University Press, 1985), 35.

33. Sam Truett, "A Wilderness of One's Own: Sports Hunting and Manliness in Nineteenth-Century America," seminar paper, Yale University Department of History, May 25, 1992.

34. John F. Reiger, *American Sportsmen and the Origins of Conservation* (1975; repr. Norman: University of Oklahoma Press, 1985); Truett, "A Wilderness of One's Own"; John M. Mackenzie, "The Imperial Pioneer and Hunter and the British Mas-

culine Stereotype in Late Victorian and Edwardian Times," in J. A. Mangan and James Walvin, eds., *Manliness and Morality: Middle-Class Masculinity in Britain and America, 1800–1940* (Manchester: Manchester University Press, 1987), 176–198.

35. Among the most insightful American authors on American game laws are James Tober, *Who Owns the Wildlife? The Political Economy of Conservation in Nineteenth-Century America* (Westport, Conn.: Greenwood Press, 1981); Stuart Marks, *Southern Hunting in Black and White: Nature, History, and Ritual in a Carolina Community* (Princeton, N.J.: Princeton University Press, 1991); Edward D. Ives, *George Magoon and the Down East Game War: History, Folklore, and the Law* (Urbana: University of Illinois Press, 1988); Steve Hahn, "Hunting, Fishing, and Foraging: Common Rights and Class Relations in the Postbellum South," *Radical History Review* 26 (1982): 37–64. Other historians have documented the rise of wildlife conservation from the perspective of administrative centers, and the elites who supported them. Reiger, *American Sportsmen and the Origins of Conservation;* James B. Trefethen, *An American Crusade for Wildlife* (New York: Winchester Press and the Boone and Crockett Club, 1975); Thomas R. Dunlap, *Saving America's Wildlife* (Princeton, N.J.: Princeton University Press, 1988).

36. Trefethen, *American Crusade,* 136.

37. For Wyoming, see *Annual Reports of the Department of the Interior for Fiscal Year Ended June 30, 1904, U.S. Office of Indian Affairs. Part I. Report of the Commissioner* (Washington, D.C., 1905), 125–128; for Montana see Dave Walter, "Swan Valley Shootout, 1908," *Montana Magazine* 69 (January–February 1985), 26–31; for New Mexico see chapter 3; for Yellowstone controversies see Karl Jacoby, "The Recreation of Nature: A Social and Environmental History of American Conservation, 1872–1918" (Ph.D. diss., Yale University, 1997), Trefethen, *American Crusade,* 89–90; for Glacier National Park, see chapter 5.

38. Feeny et al., "The Tragedy of the Commons," 4–5; also McCay and Acheson, "Human Ecology of the Commons."

39. Russell L. Robbins, Don E. Redfearn, and Charles P. Stone, "Refuges and Elk Management," 496, in Thomas and Toweill, *Elk of North America,* 479–507.

40. This is true nowhere more so than in Jackson Hole, where fights over the size of the elk herd frequently preoccupy locals and federal authorities. See Robert W. Righter, *Crucible for Conservation: The Creation of Grand Teton National Park* (Boulder: Colorado Associated University Press, 1982), 8–9, 53–54.

41. Patricia Nelson Limerick, *Legacy of Conquest: The Unbroken Past of the American West* (New York: W. W. Norton, 1987); White, *"It's Your Misfortune and None of My Own";* Cronon, *Nature's Metropolis;* also William Cronon, George Miles, and Jay Gitlin, "Becoming West: Toward a New Meaning for Western History," in William Cronon, George Miles, and Jay Gitlin, eds., *Under an Open Sky: Rethinking America's Western Past* (New York: W. W. Norton, 1992), 3–27.

42. White, *"It's Your Misfortune and None of My Own,"* 391.

43. USCIA, *Annual Report of the Commissioner of Indian Affairs 1895,* 79.

Chapter 1. The Killing of Seely Houk

1. Supreme Court of Pennsylvania, Western District, *Commonwealth vs. Rocco Racco, Appellant,* Paperbook of Appellant, No. 82, October Term, 1909, 6; Pinkerton

Reports (hereafter PR), Feb. 5, 1908; March 5, 1908, Pennsylvania Game Commision, Harrisburg. Pinkerton detectives investigating the murder sent daily reports to their superiors in Philadelphia, who sent copies to the Game Commission in Harrisburg. Copies of these reports are also on file at the Historical Society of Western Pennsylvania, Pittsburgh.

2. For examples of this treatment, see William F. Schulz, Jr., *Conservation Law and Administration: A Case Study of Law and Resource Use in Pennsylvania* (New York: Ronald Press, 1953), 45–48; *Pennsylvania Game News,* November 1950, 38; Joseph Kalbfus, *Dr. Kalbfus' Book: A Sportsman's Experiences and Impressions of East and West* (Altoona, Pa.: Times Tribune Co., 1926), 289–290.

3. For death threats, see Kalbfus, *Dr. Kalbfus' Book,* 289. There is one on file in the library of the Pennsylvania Game Commission, Harrisburg.

4. James Tober, *Who Owns the Wildlife? The Political Economy of Conservation in Nineteenth-Century America* (Westport, Conn.: Greenwood Press, 1981). See also Arthur McEvoy, *The Fisherman's Problem: Ecology and Law in the California Fisheries, 1850–1980* (New York: Cambridge University Press, 1986), 93–119, and Michael J. Bean, *The Evolution of National Wildlife Law* (New York: Praeger, 1983), 12–17. For specific legislative actions in Pennsylvania, see Commonwealth of Pennsylvania, *Digest of Game and Fish Laws 1903* (Harrisburg: State Printing Office, 1903), especially 12–13, 17–22.

5. Commonwealth of Pennsylvania, *Digest of Game and Fish Laws 1903,* 10, 26, 32.

6. PR Feb. 4, 1908; also agent J.H.G. conversation with O. L. Miller in PR Feb. 20, 1908.

7. PR Feb. 4, 1908.

8. Katherine Mayo, *Justice to All: The Story of the Pennsylvania State Police* (New York: G. P. Putnam's Sons, 1917), 64–65, 315–317.

9. Pennsylvania Board of Game Commissioners, *Report of the Game Commission of the State of Pennsylvania 1902* (Harrisburg, 1903), 4. (Hereafter all annual reports of the PBGC will be abbreviated *ARBGC*).

10. Ibid.

11. William T. Hornaday, *Our Vanishing Wildlife* (New York: Charles Scribner's Sons, 1913), esp. 94–104, 384.

12. In New York, Italian poachers were active in the environs of the New York Zoological Park (later the Bronx Zoo) soon after 1900, when employees bent on stopping illegal hunting of songbirds engaged them in armed combat. Remarkably, no one was killed. Ibid., 101–102.

13. See Joseph A. Mussulman, *Music in the Cultured Generation* (Evanston, Ill.: Northwestern University Press, 1971), 38–39; H. A. Haweis, *Music and Morals* (New York, 1874). My thanks to Clifford Doerksen for the references.

14. John F. Lacey, quoted in Col. G. O. Shields, "A Tribute to Major Lacey from a Fellow Bird Lover," in L. H. Pammel, ed., *Major John F. Lacey Memorial Volume* (Cedar Rapids: Iowa Park and Forestry Association, 1915), 17.

15. Frank M. Chapman, *The Economic Value of Birds to the State* (Albany: J. B. Lyon Company, 1903), 10.

16. *ARBGC 1909,* 9; Kalbfus, *Dr. Kalbfus' Book,* 279–280. The quotation is from the New Mexico Governor's Proclamation for Arbor and Bird Days, "Birds Need Protec-

tion as Well as Trees," *Silver City* (N. Mex.) *Independent*, March 21, 1916, 2. The Pennsylvania Game Commission's rhetoric was no less apocalyptic. See *New Castle News*, Sept. 19, 1908, pp. 1, 6. See also Chapman, *The Economic Value of Birds to the State.*

17. Hornaday, *Our Vanishing Wildlife*, 102.

18. Edward K. Muller, *A Concise Historical Atlas of Pennsylvania* (Philadelphia: Temple University Press, 1989), 89; John Higham, *Strangers in the Land: Patterns of American Nativism, 1860–1925*, 2d ed. (New York: Atheneum, 1963), 47, 66–67, 140–144; Francis A. J. Ianni, "Mafia and the Web of Kinship," in Luciano J. Iorizzo, ed., *An Inquiry into Organized Crime: Proceedings of the Third Annual Conference of the American Italian Historical Association* (Staten Island, N.Y., Oct. 24, 1970), 11.

19. Muller, *Concise Historical Atlas of Pennsylvania*, 89.

20. John Bodnar, "The Italians and Slavs of New Castle: Patterns in the New Immigration," *Western Pennsylvania Historical Magazine* 55 (1972): 269; Higham, *Strangers in the Land*, 66–67.

21. Bodnar, "The Italians and Slavs of New Castle," 271–272. For an assessment of Hillsville's ethnic demography in 1906, see Agent #37, PR April 6, 1907, "The Black Hand Society in Lawrence County, Pennsylvania, 1906," unpublished collection, vol. 1, New Castle Public Library, New Castle, Pa.

22. Hornaday, *Our Vanishing Wildlife*, 94–101.

23. PR June 15, 1906; PR June 23, 1906; PR Aug. 9, 1906; PR Aug. 10, 1906; PR Oct. 27, 1906; PR Oct. 30, 1906; PR Nov. 7, 1906; statement of Silas Martin to agent J.H.G., PR Feb. 12, 1908.

24. PR Oct. 27, 1906.

25. Frank Piscueneri, interview with the author, April 24, 1991, Hillsville, Pa. All tape recordings of interviews in possession of the author.

26. Alfred Retort, interview with the author, April 23, 1991, Hillsville, Pa.

27. Fred Iovanella, interview with the author, April 23, 1991, New Castle, Pa.

28. For a sample of the rich oral traditions regarding poachers, game wardens, locals, and outsiders in northern Pennsylvania, see James York Glimm, *Flatlanders and Ridgerunners: Folktales from the Mountains of Pennsylvania* (Pittsburgh: University of Pittsburgh Press, 1983), 4–7, 10–16.

29. *ARBGC 1908* (Harrisburg: State Printing Office, 1909), 24. For 1902 complaint, see *Report of the Game Commission of the State of Pennsylvania 1902*, 4–5. Such complaints became a constant refrain in Game Commission reports until 1910. *ARBGC 1905* (Harrisburg, 1906), 6, 10; *ARBGC 1907* (Harrisburg, 1908), 3–4; *ARBGC 1908*, 24–25; *ARBGC 1910*, 19.

30. Commonwealth of Pennsylvania, *Digest of Game and Fish Laws 1903*, 33.

31. *ARBGC 1905*, 10.

32. Cost of the license is in Commonwealth of Pennsylvania, *Digest of Game and Fish Laws 1903*, 32–33; for wages in Hillsville quarries, see "#42 Reports," PR Feb. 14, 1908. Becoming a citizen was too difficult for most immigrants in the New Castle vicinity even to consider, requiring as it did two naturalized witnesses to attest to the petitioner's character, in addition to an oral exam on American government before the often hostile Naturalization Examiner Ragsdale. See *New Castle News*, Sept. 12, 1908, p. 5.

33. David J. Cuff et al., eds., *The Atlas of Pennsylvania* (Philadelphia: Temple University Press, 1989), 52.

34. Editorial, *New Castle News,* July 15, 1907, p. 4.

35. Joe Rich, interview with author, Hillsville, Pa., April 24, 1991; Frank Piscueneri, interview with the author, Hillsville, Pa., April 24, 1991.

36. PR Jan. 23, 1908; "F.P.D. reports," in PR March 10, 1908.

37. PR Aug. 14, 1906; for discussion of the murderer's identity among local Italians, see PR Oct. 20, 1906; PR Nov. 23, 1906. Dominic Sianato was sentenced to twenty years in prison shortly before Houk's convicted murderer was hanged. See PR June 19, 1908.

38. J.H.G. interview with O. L. Miller, PR Feb. 20, 1908.

39. For Houk's clothing, see *Commonwealth vs. Rocco Racco,* 31, 47.

40. PR Feb. 4, 1908.

41. PR Feb. 4, 1908; PR Feb. 20, 1908; PR Feb. 21, 1908; "J.H.G. Reports," PR Feb. 22, 1908.

42. *ARBGC 1905,* 13.

43. *ARBGC 1907,* 15. The Game Commission claimed that all the assailants were immigrants, but whether they were is impossible to establish. The names of the other wardens who were killed and the locations of the killings have been lost, although one Italian was killed in a shootout and one of the wounded wardens identified his assailants as Italians. The Secretary of the Game Commission, Dr. Joseph Kalbfus, believed that all the shootings were perpetrated by Italian immigrants. Although this judgment sounds like ethnic prejudice, it is also true that the Non-Resident License Law was hardest on immigrants, who would have had most reason to resist. See Kalbfus, *Dr. Kalbfus' Book,* 289.

44. "Houk Guilty of Assault," *New Castle News,* Sept. 22, 1905, pp. 1, 4.

45. "Houk Guilty of Assault," *New Castle News,* Sept. 22, 1905, pp. 1, 4; *New Castle News,* March 26, 1906, p. 1; PR March 3, 1908; "J.H.G. Reports," PR March 4, 1908. See also the trial docket for the case, *Commonwealth vs. Seely Houk,* Quarter Sessions 17, September 1905, No. 23, Lawrence County Court House, New Castle, Pa.

46. PR Feb. 5, 1908; *Commonwealth vs. Rocco Racco,* 25–26, 38–39.

47. "Report of Agent #10," PR May 27, 1908; *Commonwealth vs. Rocco Racco,* 94–102.

48. "#10 Reports," PR June 5, 1908; *Commonwealth vs. Rocco Racco,* 98.

49. "Report of Agent #10," PR May 27, 1908.

50. For a comprehensive discussion of the pervasive Black Hand phenomenon, see Thomas Monroe Pitkin and Francesco Cordasco, *The Black Hand: A Chapter in Ethnic Crime* (Totowa, N.J.: Littlefield, Adams, 1977).

51. Racco testified that it was called the Society of Honor, although this may have been an attempt to dignify it; others testified that it had no real name. *Commonwealth vs. Rocco Racco,* 146–147, 226. Other Italians in the area almost certainly called the organization "Black Hand." See "The Black Hand Society in Lawrence County, 1906," unpublished collection, vols. 1 and 3, New Castle Public Library.

52. Diego Gambetta, *The Sicilian Mafia: The Business of Private Protection* (Cambridge, Mass.: Harvard University Press, 1993), 141.

53. "#89 Reports," reports of May 22, May 24, 1907, in "The Seely Houk Murder," Binder B-45, Pinkerton International Headquarters, Encino, Calif., hereafter PR-ENC.

54. "#89 Reports," PR-ENC May 24, 1907.

55. My interpretation here is drawn largely from Gambetta, *The Sicilian Mafia,* esp. 155, and Anton Blok, *The Mafia of a Sicilian Village: A Story of Violent Peasant Entrepreneurs* (New York: Harper & Row, 1974). Histories of mafia, particularly in America, usually focus on mafia after Prohibition, and therefore are of limited use for reconstructing events at the turn of the century. In the vast literature on mafia see: Ianni, "Mafia and the Web of Kinship," 1–22; Pino Arlacchi, *Mafia Business: Mafia Ethic and the Spirit of Capitalism,* tr. Martin Ryle (London: Verso, 1986); William Balsamo and George Carpozi, Jr., *Under the Clock: The Inside Story of the Mafia's First Hundred Years* (Far Hills, N.J.: New Horizon Press, 1988); Stephen R. Fox, *Blood and Power: Organized Crime in Twentieth-Century America* (New York: William Morrow, 1989).

56. For Racco's family as central, see "Supt. AAE Reports," PR June 12, 1906.

57. PR Jan. 13, 1907; PR Feb. 11, 1907. Apparently the exam involved secret codewords: PR Jan. 8, 1907.

58. PR Oct. 13, 1906; PR Oct. 14, 1906.

59. For guaranteed board payments see PR Jan. 16, 1907; for credit management see PR Sept. 3, 1906; for sickness, injury, and other compensation see PR Oct. 15, 1906; PR Oct. 16, 1906; PR Jan. 9, 1907; PR April 6, 1907.

60. See Blok, *The Mafia of a Sicilian Village,* 154, and Candido's interview with detectives, PR March 28, 1908.

61. "Agent #37 Reports," April 6, 1907, in "The Black Hand Society in Lawrence County, 1906," vol. 1.

62. PR July 16, 1906; PR Aug. 25, 1906; Gambetta, *The Sicilian Mafia,* 48–52.

63. PR Oct. 21, 1906.

64. Blok, *The Mafia of a Sicilian Village.*

65. For Racco's store, see Candido's interview with FPD, PR March 28, 1908; for bootlegging, see PR July 3, 1906; PR July 15, 1906; PR July 9, 1906 (for some reason, there are two separate reports by D.P. for this date; only one mentions beer selling); PR Dec. 29, 1906.

66. PR Feb. 19, 1907.

67. PR Aug. 6, 1906. The society took special interest in controlling violence as it came under increasing pressure from detectives; see PR Sept. 12, 1906; PR Sept. 19, 1906.

68. PR July 18, 1906.

69. I have found no evidence of any female members of the society.

70. See *New Castle News,* July 23, 1907, pp. 1, 6.

71. PR April 21, 1908.

72. PR Feb. 17, 1908.

73. PR Oct. 15, 1906; PR Feb. 21, 1908.

74. PR June 5, 1908.

75. *Commonwealth vs. Rocco Racco,* 227–228, 273.

76. Gambetta, *The Sicilian Mafia,* 93.

77. *Commonwealth vs. Rocco Racco,* 222–229; *New Castle News,* Sept. 18, 1908, pp. 1, 4, 10, covers Racco's testimony at his murder trial, which is confirmed by subsequent witnesses and the following sources. The strike itself was covered in *New Castle News,* Feb. 6, 1906, p. 1. For the calculated mendacity of the adultery charge, see FPD's report of interview with Salvatore Candido, PR March 28, 1908; also, PR Jan. 20, 1907, indicates that "improper" sexual relations were within the society's purview of so-

cial regulation. See also PR Feb. 21, 1907. Racco claimed he and Surace had fallen out over the strike, the adultery charge, and division of extortion spoils. *New Castle News,* Sept. 19, 1908, p. 10.

78. The society headman was Joe Bagnato. PR Sept. 19, 1906; *New Castle News,* Sept. 21, 1908, p. 2. Frank Piscueneri recalls that his father would often say "when I was born—nineteen six—that was rough days. 'Five and 'six, them two years up here, was bad" (interview with the author).

79. PR March 28, 1908; *New Castle News,* Sept. 18, 1908, pp. 1, 4; *New Castle News,* Sept. 19, 1908, p. 10.

80. *Commonwealth vs. Rocco Racco,* 113–114; "#10 Reports," PR May 25, 1908; PR May 28, 1908; PR June 3, 1908.

81. Testimony of Herman List, in *Commonwealth vs. Rocco Racco,* 183–184.

82. "#10 Reports," PR May 28, 1908; see the testimony of Ferdinand Serace, *Commonwealth vs. Rocco Racco,* 110–120.

83. Testimony of Joe Cutrone, in *Commonwealth vs. Rocco Racco,* 167.

84. PR Feb. 15, 1908; PR May 25, 1908.

85. PR March 10, 1908; "Supt. FPD reports"; Report of Agent #37, April 6, 1907, "The Black Hand Society in Lawrence County, 1906," vol. 1.

86. "#89 Reports," PR-ENC, June 14, 1907.

87. *New Castle News,* Sept. 19, 1908, pp. 1, 6.

88. *New Castle News,* Oct. 26, 1909, p. 1.

89. PR Sept. 22, 1906.

90. Statement of Nora Martin to J.H.G., PR Feb. 13, 1908; Statement of Silas Martin to J.H.G., PR Feb. 12, 1908; PR Feb. 5, 1908; PR Feb. 24, 1908.

91. "Bagnato Gone; So Has $11,000 Italians' Gold," *New Castle News,* July 13, 1907.

92. PR Oct. 19, 1906; PR Feb. 6, 1907; PR Feb. 28, 1907.

93. *New Castle News,* Aug. 10, 1907, p. 2.

94. In 1907, a number of game wardens were shot at from a distance, "more to notify the officer that the pursued was armed and to thus intimidate him, than to injure or to kill." *ARBGC 1907,* 15.

95. "Five Arrests in Hillsville," *New Castle News,* Oct. 26, 1909, p. 1; "Five Paid Fines; Three Are Jailed," *New Castle News,* Oct. 27, 1909, p. 1.

96. *ARBGC 1910,* 19.

97. "#10 Reports," PR May 27, 1908.

98. The Game Commission reported 384 prosecutions for violations of the Alien Gun Law in 1927–28, 247 in 1928–29, and 228 in 1931–32. See *Biennial Report of the Game Commissioners, 1927–28* (title varies; hereafter *BRGC*); *BRGC, 1927–28,* 10; *BRGC, 1929–30; BRGC, 1931–32,* 8.

99. Joe Rich, interview with author.

100. Fred Iovanella, interview with author.

101. Field Notes, John Rich, interview with author, Hillsville, Pa., April 24, 1991.

Chapter 2. Boon and Bust

1. William F. Schulz, Jr., *Conservation Law and Administration: A Case Study of Law and Resource Use in Pennsylvania* (New York: Ronald Press, 1953), 47.

2. Ibid.; Michael Sajna, "The Forgotten Savior. . . . John M. Phillips," *Pennsylvania Game News*, March 1990, 11–15.

3. Pennsylvania Board of Game Commissioners, *Biennial Report of the Board of Game Commissioners, 1927–28*, 31–32. Hereafter *BRBGC*.

4. Schulz, *Conservation Law and Administration*, 25.

5. Samuel P. Hays, *Conservation and the Gospel of Efficiency: The Progressive Conservation Movement* (Cambridge, Mass.: Harvard University Press, 1959); Steve Hahn, "Hunting, Fishing, and Foraging: Common Rights and Class Relations in the Postbellum South," *Radical History Review* 26 (1982): 37–64.

6. Joseph Kalbfus, *Dr. Kalbfus' Book: A Sportsman's Experiences and Impressions of East and West* (Altoona, Pa.: Times Tribune Co., 1926), 93.

7. J. Q. Creveling, "Conservation in Pennsylvania: Fifty Years of Progress," n.d., Pennsylvania Game Commission, Harrisburg.

8. Henry W. Shoemaker, *Pennsylvania Deer and Their Horns* (Reading, Pa.: Faust Printing Co., 1915); George H. Matschke et al., "Population Influences," in Lowell K. Halls, ed., *White-Tailed Deer: Ecology and Management* (Harrisburg, Pa.: Stackpole Books, 1984), 171.

9. Shoemaker, *Pennsylvania Deer and Their Horns*, 26.

10. Edward K. Muller, ed., *A Concise Historical Atlas of Pennsylvania* (Philadelphia: Temple University Press, 1989), 95.

11. Richard F. Harlow, "Habitat Evaluation," in Halls, ed., *White-Tailed Deer*, 606–607.

12. Muller, *Concise Historical Atlas of Pennsylvania*, 95.

13. Ibid.

14. Wallace Byron Grange, *The Way to Game Abundance* (New York: Scribner, 1949), 256; Harlow, "Habitat Evaluation," 625; Matschke et al., "Population Influences," 186.

15. Shoemaker, *Pennsylvania Deer and Their Horns*, 9.

16. *Report of the Game Commission of the State of Pennsylvania 1902*, 3; Commonwealth of Pennsylvania, *Digest of the Game and Fish Laws and Warden and Forestry Laws 1903* (Harrisburg, 1903), 19. "Jack lighting" was the practice of hunting deer at night, shining lights into the animals' eyes to stun them so they could be more easily killed.

17. For an explanation of secondary growth on clear-cuts, see F. H. Bormann and G. E. Likens, *Pattern and Process in a Forested Ecosystem* (New York: Springer-Verlag, 1979), 103–137. Specifics of this phenomenon in Pennsylvania are discussed below.

18. Harlow, "Habitat Evaluation," 601, 606.

19. *Harrisburg Courier*, Dec. 22, 1907, section 2, p. 1. Clippings from this source are in Miscellaneous Historical Files, Pennsylvania Game Commission Library, Harrisburg.

20. Pennsylvania Board of Game Commissioners, *Annual Report of the Board of Game Commissioners 1911* (hereafter *ARBGC*), 11.

21. *Hunter-Trader-Trapper* 16, no. 1 (April 1908): 113.

22. *ARBGC 1906*, 9–10; *ARBGC 1911*, 14–15.

23. *ARBGC 1911*, 14.

24. *Harrisburg Courier*, Dec. 22, 1907, section 2, p. 1.

25. Kenneth E. F. Watt, *Ecology and Resource Management: A Quantitative Approach* (New York: McGraw-Hill, 1968), 128; Grange, *The Way to Game Abundance*,

180; Matschke et al., "Population Influences," 170; Dale R. McCullough, "Lessons from the George Reserve, Michigan," in Halls, ed., *White-Tailed Deer,* 231.

26. *ARBGC 1911,* 10.

27. "Game Dealers Are Caught," *In the Open,* 1, no. 6 (February 1912): 45.

28. Ibid.

29. "Enforce the Laws," *In the Open* 1, no. 8 (April 1912): 28.

30. "Game Dealers Are Caught," 44.

31. "More Wardens Are Needed," *In the Open,* 1, no. 4 (November 1911): 34.

32. Ibid.

33. "Game Dealers Are Caught," 45.

34. *ARBGC 1913,* 7–8.

35. Schulz, *Conservation Law and Administration,* 48, 86.

36. Schulz, *Conservation Law and Administration,* 60.

37. My interpretation of hunting licenses as bills of sale takes issue with the scholars of resource policy who see more rigid divisions between bureaucracy and markets. See, for example, John Baden and Richard Stroup, eds., *Bureaucracy vs. Environment: The Environmental Costs of Bureaucratic Governance* (Ann Arbor: University of Michigan Press, 1981).

38. *ARBGC 1916,* 30.

39. *ARBGC 1915,* 7.

40. *ARBGC 1913,* 8.

41. See *ARBGC 1915,* 36–37.

42. Schulz, *Conservation Law and Administration,* 50–51; *ARBGC 1913,* 7.

43. The amount of crop damage by foraging deer is directly proportional to deer density; the more deer, the more damage local farms will sustain. See George H. Matschke, David S. deCalesta, and John D. Harder, "Crop Damage and Control," in Halls, ed., *White-Tailed Deer,* 648.

44. *ARBGC 1915,* 35.

45. *ARBGC 1915,* 36.

46. Figures for buck kill are from Roger M. Latham, "Pennsylvania's Deer Problem," *Pennsylvania Game News* Special Issue No. 1 (September 1950): 12. These figures do not indicate the illegal take, which was no doubt quite large. Figures for license sales are from *BRBGC 1922–24,* 32.

47. For the composition of intermediate forest stands, see *BRBGC 1927–28,* 28. See also James York Glimm, *Flatlanders and Ridgerunners: Folktales from the Mountains of Northern Pennsylvania* (Pittsburgh: University of Pittsburgh Press, 1983), xxx; for the lack of deer forage in intermediate succession stages on clear-cuts, see Harlow, "Habitat Evaluation," 601; for how this condition created starvation conditions for Pennsylvania's deer, see Latham, "Pennsylvania's Deer Problem," 13–14.

48. Matschke et al. "Crop Damage and Control," 648.

49. *BRBGC 1922–24,* 7.

50. Ibid., 7–8.

51. Seth Gordon, "Conservation Madness," *Country Gentleman* 107, no. 5 (May 1937): 16; *BRBGC 1922–24,* 23–24; *BRBGC 1927–28,* 32.

52. *BRBGC 1922–24,* 8.

53. Schulz, *Conservation Law and Administration,* 52.

54. Gordon, "Conservation Madness," 17.

55. *ARBGC 1913,* 14.

56. Gordon, "Conservation Madness," 16.

57. *Pennsylvania County Reports* (Philadelphia: Geo. T. Bisel Co., 1918), *Commonwealth vs. Carbaugh,* vol. 45, 67.

58. *District and County Reports* (Philadelphia: The Legal Intelligencer, 1923), *Commonwealth vs. Gilbert,* vol. 5, 445.

59. Gordon, "Conservation Madness," 16.

60. Commonwealth of Pennsylvania, *Digest of the Game and Fish Laws and Warden and Forestry Laws 1903,* 27.

61. Schulz, *Conservation Law and Administration,* 52.

62. *BRBGC 1927–28,* 38.

63. *BRBGC 1927–28,* 37.

64. Schulz, *Conservation Law and Administration,* 52.

65. *BRBGC 1927–28,* 29.

66. Ibid., 32. The game commission did report mysterious declines in rabbit populations in the early twenties. See *BRBGC 1922–24,* 28–29.

67. Ibid., 33.

68. Matschke et al., "Crop Damage and Control," 170–171; McCullough, "Lessons from the George Reserve," 231. Fawns would starve in larger numbers because (1) as Vernon Bailey observed in his survey, in heavily browsed forests, fawns are not tall enough to reach the "deer line," and (2) young animals require more calories than adults for each pound of body weight. For the second point, see Leonard Wing, *The Practice of Wildlife Conservation* (New York: Wiley, 1951), 23.

69. See for example Pennsylvania Board of Game Commissioners, *More Food for Upland Game,* Bulletin No. 11 (Harrisburg, n.d.).

70. *BRBGC 1927–28,* 31; Gordon, "Conservation Madness," 17.

71. Schulz, *Conservation Law and Administration,* 52.

72. Latham, "Pennsylvania's Deer Problem."

73. Latham, "Pennsylvania's Deer Problem," 21.

74. Creveling, "Conservation in Pennsylvania," 4. Creveling's manuscript is not dated, but I calculate the date to be roughly 1950 from comparing the subtitle "Fifty Years of Progress" to the date of the Game Commission's creation, 1895.

Chapter 3. "Raiding Devils" and Democratic Freedoms

1. Curt Meine, *Aldo Leopold: His Life and Work* (Madison: University of Wisconsin Press, 1988), 152.

2. Aldo Leopold, *Game Management* (New York: C. Scribner's Sons, 1933).

3. Aldo Leopold, *A Sand County Almanac* (New York: Oxford University Press, 1949).

4. Paul Hirt, *A Conspiracy of Optimism: Management of the National Forests Since World War II* (Lincoln: University of Nebraska Press, 1994), 61.

5. For more complete treatments of the Maxwell Land Grant's tumultuous history, see William Keleher, *Maxwell Land Grant: A New Mexico Item* (1942; repr. New York: Argosy-Antiquarian, 1964); Jim Berry Pearson, *The Maxwell Land Grant* (Norman: University of Oklahoma Press, 1961); Maria E. Montoya, "Dispossessed People:

Settler Resistance on the Maxwell Land Grant, 1860–1901" (Ph.D. diss., Yale University, 1993).

6. Keleher, *Maxwell Land Grant*, 1–44; Pearson, *The Maxwell Land Grant*, 40–54, 62; Montoya, "Dispossessed People," 173.

7. Howard R. Lamar, *The Far Southwest, 1846–1912: A Territorial History* (1966; repr. New York: W. W. Norton, 1970), 147.

8. Ibid., 496.

9. Schomburg to Bartlett, March 7, 1901, William Henry Bartlett Papers, Center for Southwest Research, University of New Mexico (hereafter Bartlett Papers); Anon., "Page B. Otero: New Mexico's Pioneer Game Warden," *New Mexico Conservationist* 2, no. 3 (April 1929): 4.

10. "William H. Bartlett," *Price Current-Grain Reporter*, Dec. 18, 1918, 3, clipping in index file, Bartlett Papers.

11. They also secured regular rail service over the mountains to deliver mail at Bartlett's post office. Pearson, *The Maxwell Land Grant*, 231.

12. "William H. Bartlett," 3.

13. It appears that there was opposition to the reservation of landowner prerogatives, notably from Governor Otero himself, but the statute passed anyway. Miguel Otero to G. O. Shields, Dec. 18, 1902, Territorial Archives of New Mexico (hereafter TANM), reel 143, frame 613.

14. *Acts of the Legislative Assembly of the Territory of New Mexico, 35th Session, 1903* (Santa Fe: New Mexican Printing Co., 1903), 79.

15. *Acts of the Legislative Assembly of the Territory of New Mexico, 36th Session, 1905* (Santa Fe: New Mexican Printing Co., 1905), 102.

16. Territory of New Mexico, "Report of the Territorial Game and Fish Warden, 1909," TANM, reel 177, frame 621.

17. Socorro County was second, with twelve, the number of a party of Navajos arrested for hunting deer out of season and exceeding the bag limit that year. Other counties had four or fewer arrests. Three counties had only one arrest each. Although there was no breakdown of the arrests by violation, Griffin provided a breakdown of arrests by violation across the territory. The number-one offense was trespassing on private property, with fourteen arrests, a category which seems likely to have been enforced on the Maxwell Land Grant. Ibid., frame 623.

18. Costs of park licenses were $15 for two years, $50 for ten years. In addition, the language of the 1912 park law suggests that such parks were effectively already in existence, since it concerned "Any person *having already established or desiring to establish* a park or lake for the purpose of keeping or propagating and selling the game or game fish therein" (emphasis added). *Laws of the State of New Mexico, 1912* (Albuquerque: Albright and Anderson, 1912), 261–267.

19. In fact, some of the first private game parks, in the latter nineteenth century, were in Pennsylvania. Notable was the Blooming Grove Park Association in Pike County, founded in 1871. In addition, there were numerous private hunting clubs throughout New York's Adirondacks by the latter nineteenth century. See James A. Tober, *Who Owns the Wildlife? The Political Economy of Conservation in the Nineteenth Century* (Westport, Conn.: Greenwood Press, 1981), 126–128.

20. Figure for New Mexico is from New Mexico Game and Fish Warden (hereaf-

ter NMGFW), *Biennial Report of the Game and Fish Warden of the State of New Mexico, 1917–1918* (Santa Fe, 1919), 62. For Pennsylvania figure, see William F. Schulz, Jr., *Conservation Law and Administration: A Case Study of Law and Resource Use in Pennsylvania* (New York: Ronald Press, 1953), 86.

21. Robert D. Baker, Robert S. Maxwell, Victor H. Treat, and Henry Dethloff, *Timeless Heritage: A History of the Forest Service in the Southwest* (College Station, Tex.: USDA, 1988), 25–26.

22. NMGFW, *First Report of Game and Fish Warden for New Mexico, 1909–1911* (Santa Fe, 1912), 17.

23. NMGFW, *Report of the Game and Fish Warden, 1912–1914* (Santa Fe, 1915), 23.

24. NMGFW, *Biennial Report of the Game and Fish Warden of the State of New Mexico 1917–1918,* 37–39.

25. Ibid., 39.

26. Ibid.

27. For a discussion of the role of tourism in early efforts to restock New Mexico's streams with trout, see Minium to Otero, Feb. 20, 1898, TANM, reel 129, frame 135. For tourism in game protection, see NMGFW, "Annual Report to the Governor" (Santa Fe: July 10, 1905), in Elliott Barker Papers, box 21, uncat. collection, Rio Grande Historical Collections, New Mexico State University, Las Cruces; see also NMGFW, "Annual Report to the Governor," July 6, 1904, same location. At times, the Game and Fish Department took on the role of advertising agency for the territory's recreational opportunities. See Otero to Outdoor Life Publishing Company, May 23, 1905, *Letterbook 1905–6,* 43, New Mexico Game and Fish Records, file 77, New Mexico State Records Center and Archives, Santa Fe (hereafter *Letterbook 1905–6*).

28. Gable's history suggests that he was an associate of the Santa Fe Ring before his appointment as game warden. He was a personal friend of Governor George Curry, having bailed Curry out of jail during the Colfax County War in the 1880s, and contributed money for Curry's trip to Washington to lobby for appointment as territorial governor in 1885. He was also "an enthusiastic sportsman." H. B. Hening, ed., *George Curry, 1861–1947: An Autobiography* (Albuquerque: University of New Mexico Press, 1958), 50, 52, 55, 204, 232, 235. Throughout Gable's career, connections to powerful politicians brought him plum appointments in the territorial and state governments: before Governor Curry appointed him territorial game warden in 1909, he had been warden of the New Mexico penitentiary and postmaster of Santa Fe. C. S. Petersen, ed., *Representative New Mexicans 1912* (Denver: C. S. Petersen, 1912), 108. For purchase of elk, see clipping from *Las Vegas Optic,* May 18, 1911, in "Charges by J. H. Sloan against Thomas P. Gable, Territorial Game and Fish Warden, for mismanagement," TANM, reel 188, frame 1310. The town of Dawson, a Colfax County coal town, was named for J. B. Dawson. T. M. Pearce, ed., *New Mexico Place Names: A Geographical Dictionary* (Albuquerque: University of New Mexico Press, 1965), 45.

29. NMGFW, "First Report of Game and Fish Warden for New Mexico," 48–49, NMDGF.

30. J. H. Sloan to W. J. Mills, TANM, reel 188, frame 1308.

31. Quoted in Meine, *Aldo Leopold,* 169. American game law had developed partly in opposition to European game laws, which held that landowners owned the

game on their estates. Tober, *Who Owns the Wildlife?* 146–148; Thomas A. Lund, *American Wildlife Law* (Berkeley: University of California, 1980), 20–27.

32. Aldo Leopold, "The National Forests: The Last Free Hunting Grounds of the Nation," *Journal of Forestry* 17, no. 2 (1919): 150–153.

33. Meine, *Aldo Leopold,* 161.

34. NMGFW, "Report of the Territorial Game and Fish Warden," 1909, TANM, reel 177, frame 626.

35. NMGFW, *Report of the Game and Fish Warden of State of New Mexico 1917–1918,* 65.

36. By far the best discussion of the fight between homesteaders and grant owners is Montoya, "Dispossessed People," esp. 222–256.

37. Van Houten to Barker, Sept. 17, 1932, MSS 147, Maxwell Land Grant Papers, item 115, Center for Southwest Research, University of New Mexico.

38. J. Stokley Ligon, "Game Ranges and Refuges," *New Mexico Magazine,* October 1932, 27.

39. NMGFW, *Report of the Game and Fish Warden of State of New Mexico 1917–1918,* 38.

40. New Mexico Game Protective Association, *Game Protective Association News (Las Cruces),* Jan. 24, 1936, 3, in Record Group 95, Records of the Forest Service, Division of Wildlife Management, 1914–1950, W. Cooperation, Box 3, Folder "W. Cooperation R-3 State of New Mexico, 1926–1936," National Archives, Suitland, Md.

41. New Mexico Game Protective Association, "Proceedings of the Convention," Albuquerque, March 10–11, 1916, RG 95, Records of the Forest Service, Division of Wildlife Management, General Correspondence, W. Cooperation, Region 3, Box 3, Folder "W. Cooperation R-3 General 1915–36." For a description of the state organization's creation and its goals, see Meine, *Aldo Leopold,* 154.

42. Meine, *Aldo Leopold,* 152.

43. Lamar, *The Far Southwest,* 487–488.

44. U.S. Dept. of Agriculture, Forest Service, Southwestern Region, *Cultural Resources Management: The Early Days,* Book 2, Report No. 11, March 1991, 26.

45. Otero to Perry, Oct. 31, 1905, *Letterbook 1905–6,* 153.

46. Ibid.

47. Borrowdale to Otero (copy), Nov. 6, 1905, *Letterbook 1905–6.*

48. Otero to Department of Interior, Nov. 25, 1905, *Letterbook 1905–6,* 228; quotations from Otero to Fullerton, Nov. 8, 1905, *Letterbook 1905–6,* 185.

49. Richard L. Nostrand, "The Hispano Homeland in 1900," *Annals of the Association of American Geographers* 70, no. 3 (September 1980): 384.

50. All population figures are from Richard L. Nostrand, *The Hispano Homeland* (Norman: University of Oklahoma Press, 1992), 60, 67.

51. Stella Hughes, *Hashknife Cowboy: Recollections of Mack Hughes* (Tucson: University of Arizona, 1984), 20; Agnes Morley Cleaveland, *No Life for a Lady* (1941; repr. Lincoln: University of Nebraska Press, 1977), 23–25.

52. Vernon Bailey, *Mammals of New Mexico* (Washington, D.C.: Government Printing Office, 1931), 32, 36.

53. Edward F. Castetter and Morris E. Opler, "The Ethnobiology of the Chiricahua and Mescalero Apache: A. The Use of Plants for Foods, Beverages and Narcotics,"

University of New Mexico Bulletin, Ethnobiological Studies in the American Southwest 3, Biological Series 4 (5), 1936, 4–5; Thomas D. Hall, *Social Change in the Southwest, 1350–1880* (Lawrence: University Press of Kansas, 1989), 228–230.

54. Vernon Bailey, *Life Zones and Crop Zones of New Mexico* (Washington, D.C.: Government Printing Office, 1913), 29; Peggy A. Gerow et al., *The Boyd Land Exchange Survey: A Cultural Resources Inventory of Public Lands in West-Central New Mexico* (Albuquerque: University of New Mexico Office of Contract Archaeology, 1994), 32–35; T. J. Ferguson and E. Richard Hart, *A Zuni Atlas* (Norman: University of Oklahoma, 1985), 43–44; Dr. Robert L. Rands, *Laguna Land Utilization: An Ethnohistorical Report,* Indian Claims Commission Docket 227, *Pueblo de Laguna vs. United States of America,* n.d., p. 281; Frank D. Reeve, "The Navajo Indians," in Myra Ellen Jenkins, ed., *Navajo Indians II* (New York: Garland, 1974), 85–98; Richard F. Van Valkenburgh, "A Short History of the Navajo People," 278–279, in David Agee Horr, ed., *Navajo Indians III* (New York: Garland, 1974), 201–268; Morris E. Opler, "Chiricahua Apache," in Alfonso Ortiz, ed., *Handbook of North American Indians* (Washington, D.C.: Smithsonian Institution, 1983), 10, 411–413. I am borrowing from the "buffer zone" concept of Harold Hickerson, "The Virginia Deer and Intertribal Buffer Zones in the Upper Mississippi Valley," in Anthony Leeds and Andrew P. Vayda, eds., *Man, Culture, and Animals: The Role of Animals in Human Ecological Adjustments* (Washington, D.C.: American Association for the Advancement of Science, 1965), 43–65; see also Richard White, *The Roots of Dependency* (Lincoln: University of Nebraska Press, 1983), 66.

55. William deBuys, *Enchantment and Exploitation: The Life and Hard Times of a New Mexico Mountain Range* (Albuquerque: University of New Mexico Press, 1985), 280–284.

56. By 1890, Rio Arriba, Santa Fe, Mora, and Taos counties, four of the counties in the Sangre de Cristos, had a combined population of 37,000, the vast majority of which was dispersed in smaller settlements over the mountains. U.S. Bureau of the Census, *Twelfth Census of the United States, Bulletin No. 37* "Population of New Mexico by Counties and Minor Civil Divisions," Jan. 21, 1901, 4–5. The number of eastern Chiricahuas probably never exceeded 3,000 and in any case declined to fewer than that by the latter nineteenth century. Opler, "Chiricahua Apache," 410–411. Other Indians used the area mostly on a seasonal basis. See note 53.

57. In predominantly Hispano counties, stocking rates for sheep vastly exceeded those in predominantly non-Hispano counties. In 1880, Santa Fe, San Miguel, Rio Arriba, and Mora counties all boasted more than 20 sheep per square mile; in Valencia County there were 38, and the highest count was in Bernalillo County, where there were 71. Alvar Ward Carlson, "New Mexico's Sheep Industry, 1850–1900: Its Role in the History of the Territory," *New Mexico Historical Review* 44 (1969): 33, 38–39.

58. See Benjamin Kemp's testimony in U.S. Dept. of Agriculture, *Cultural Resources Management, The Early Days,* Book 2, 38–39.

59. In addition, most of the sheep in Socorro County were north of the Mogollons, on the Plains of San Augustin, or well to the east, along the Rio Grande. Carlson, "New Mexico's Sheep Industry," 33, 38.

60. DeBuys, 220–223. By the turn of the century, the Pecos Forest Reserve was so overgrazed that sheep were stripping the trees of their foliage. William J. Parish, *The Charles Ilfeld Company: A Study of the Rise and Decline of Mercantile Capitalism in New*

Mexico (Cambridge, Mass.: Harvard University Press, 1961), 178. For deer-livestock relationships, see Richard J. Mackie, "Interspecific Relationships," in Olof C. Wallmo, ed., *Mule and Black-Tailed Deer of North America* (Lincoln: University of Nebraska Press, 1981), 499–500. In 1913, Vernon Bailey remarked that deer in the Sangre de Cristo range were "becoming scarce." Bailey, *Mammals of New Mexico,* 56. J. Stokley Ligon explored the implications of harsh winters and heavy stock grazing in the Sangres in the mid-1920s. See J. Stokley Ligon, *Wild Life of New Mexico: Its Conservation and Management* (Santa Fe: State Game Commission, 1927), 73.

61. Ligon, *Wild Life of New Mexico,* 75. Also, Jerry L. Williams, ed., *New Mexico in Maps,* 2nd ed. (Albuquerque: University of New Mexico Press, 1986), 38–39, 52–53.

62. In 1914, three large national forests on the Sangres—the Pecos, the Carson, and the Jemez forests—produced a combined total deerkill of fewer than 50. That same year, in the Datil and Gila national forests 843 deer were killed. These numbers are problematic. We can be sure that in both places hunters took deer that they did not report, particularly in the Sangre de Cristos, where villagers lived far from forest service authorities; of course, the same could be said of locals in the Mogollons at that time. Yet legal hunters in the Mogollons took eighteen times more deer than legal hunters did in the Sangre de Cristos, and such an order of magnitude is difficult to dismiss. The 1914 hunt was no aberration. In 1915, hunters on the Santa Fe and Carson in the north forests took 48 deer. On the Gila and Datil forests, hunters took 503. Bailey, *Mammals of New Mexico,* 31–32; also Ligon, *Wild Life of New Mexico,* 73–75.

63. James McKenna, *Black Range Tales* (1936; repr. Chicago: Rio Grande Press, 1965), 50–57; Bailey, *Mammals of New Mexico,* 32, 36–37.

64. I shall cite individual articles and books as necessary, but for the discussion which follows I draw generally on these publications: Merton Leland Miller, *A Preliminary Study of the Pueblo of Taos, New Mexico* (Chicago: University of Chicago Press, 1898); Elsie Clews Parsons, "Taos Pueblo," *General Series in Anthropology* No. 2 (Menasha, Wisc.: George Banta, 1936); Leslie A. White, "The Pueblo of Santa Ana, New Mexico," *Memoirs of the American Anthropological Association,* 1942, no. 60; White, "The Pueblo of San Felipe," *Memoirs of the American Anthropological Association,* 1938, no. 38; White, "Notes on the Ethnozoology of the Keresan Pueblo Indians," *Papers of the Michigan Academy of Science Arts and Letters* (Ann Arbor: University of Michigan Press, 1947), 223–243; Junius Henderson and John Peabody Harrington, "Ethnozoology of the Tewa Indians," 1914, no. 56, *Bulletin of the Bureau of American Ethnology;* Charles H. Lange, "Cultural Change as Revealed in Cochiti Pueblo Hunting Customs," *Texas Journal of Science* (June 1953): 178–184; Esther S. Goldfrank, "Notes on Deer-Hunting Practices at Laguna Pueblo, New Mexico," *Texas Journal of Science* (1954), no. 4: 407–421; Frances H. Elmore, "The Deer and His Importance to the Navajo," *El Palacio* 60 (1953): 371–384; W. W. Hill, "The Agricultural and Hunting Methods of the Navaho Indians," *Yale University Publications in Anthropology,* 1938, no. 18; Morris E. Opler, *An Apache Life-Way* (1941; repr. Chicago: University of Chicago Press, 1965); Morris E. Opler and Edward F. Castetter, "The Ethnobiology of the Chiricahua and Mescalero Apache," *University of New Mexico Bulletin,* Ethnobiological Studies in the American Southwest, Biological Series 4 (5), 1936. For pueblo Indian hunting, a good general source is Hamilton A. Tyler, *Pueblo Animals and Myths* (Norman: University of Oklahoma Press, 1975); also useful are the relevant sections of Elsie Clews Parsons, *Pueblo Indian Religion,* 2 vols. (Chicago: University of Chicago Press, 1939).

65. White, "Pueblo of Santa Ana," 284; Hill, "The Agricultural and Hunting Methods of the Navaho Indians," 98; Opler and Castetter, "The Ethnobiology of the Chiricahua and Mescalero Apache," 18; White, "Notes on the Ethnozoology of the Keresan Pueblo Indians," 231.

66. White, "Pueblo of Santa Ana," 280–281, and "Notes on the Ethnozoology of the Keresan Pueblo Indians," 231; Goldfrank, "Notes on Deer-Hunting Practices at Laguna Pueblo," 415; Parsons, *Pueblo Indian Religion,* 1:127. For hunt society controls, see White, "Pueblo of Santa Ana," 287–289; Goldfrank, 409.

67. White, "Pueblo of Santa Ana," 294.

68. Parsons, *Pueblo Indian Religion,* 1:84.

69. White, "Pueblo of Santa Ana," 293; Parsons, "Taos Pueblo," 20.

70. Goldfrank, "Notes on Deer-Hunting Practices at Laguna Pueblo," 409.

71. White, "Pueblo of Santa Ana," 289; Goldfrank, "Notes on Deer-Hunting Practices at Laguna Pueblo," 418.

72. Hill, "The Agricultural and Hunting Methods of the Navaho Indians," 112.

73. Opler, *Apache Life-Way,* 322.

74. D. W. Meinig, *Southwest: Three Peoples in Geographical Change, 1600–1970* (New York: Oxford University Press, 1971), 55.

75. Miller, *Preliminary Study of the Pueblo of Taos,* 23; Parsons "Taos Pueblo," 19–20; Opler, *Apache Life-Way,* 316, 329; Elmore, "The Deer and His Importance to the Navajo," 375–384.

76. In the latter decades of the nineteenth century, hunters from Cochiti and other pueblos hunted the upper drainage of the Pecos River, where they often did battle with Apaches, Kiowas, and Comanches, all of whom claimed the same hunting grounds as their own. Lange, "Cultural Change as Revealed in Cochiti Pueblo Hunting Customs," 179.

77. Also, Indians regularly hunted lands in the immediate vicinity of their pueblos and reservations. In 1905, for example, Jemez Indians hunted in the Mogollons, but they were also hunting closer to the pueblo in Sandoval County. Otero to Blake, Dec. 26, 1905, *Letterbook 1905–6.*

78. Territory of New Mexico, *Acts of the Legislative Assembly of the Territory of New Mexico, 31st Session, 1895* (Santa Fe: New Mexican Printing Co., 1895), 58–60.

79. Territory of New Mexico, *Acts of the Legislative Assembly of the Territory of New Mexico, 34th Session, 1901* (Albuquerque: Democrat Publishing Co., 1901), 97; Territory of New Mexico, *Acts of the Legislative Assembly of the Territory of New Mexico, 35th Session, 1903* (Santa Fe: New Mexican Printing Co., 1903), 75.

80. Territory of New Mexico, *Acts of the Legislative Assembly of the Territory of New Mexico, 36th Session, 1905* (Santa Fe: New Mexican Printing Co., 1905), 100.

81. For an insightful discussion of the cultural disparities between conservationist and indigenous concepts of resource use, see Robert A. Brightman, "Conservation and Resource Depletion: The Case of the Boreal Forest Algonquins," in Bonnie McCay and James M. Acheson, eds., *The Question of the Commons: The Culture and Ecology of Communal Resources* (Tucson: University of Arizona Press, 1987), 130.

82. Nostrand, *The Hispano Homeland,* 75, 95, 113–115.

83. U.S. Bureau of the Census, *Twelfth Census of the United States, Bulletin No. 37,* "Population of New Mexico by Counties and Minor Civil Divisions," Washington, D.C., Jan. 21, 1901, 2.

84. Bruce Ashcroft, *The Territorial History of Socorro, New Mexico* (El Paso: Texas Western Press, 1988), 47.

85. E. O. Wooton, "The Range Problem in New Mexico," *Bulletin No. 66,* New Mexico Agricultural Experiment Station, April 1908, 27.

86. Bailey, *Mammals of New Mexico,* 25.

87. N. Hollister, "New Mexico: Datil. Mammals. Oct. 4–24, 1905," Bureau of Biological Survey Reports, History of the Forest Service in the Southwest, Box 1, Folder 17, University of New Mexico, Center for Southwest Research, Albuquerque.

88. Hughes, *Hashknife Cowboy,* 20; Cleaveland, *No Life for a Lady,* 25.

89. Interestingly, Hotchkiss also said this hunting party included women. Bailey, *Mammals of New Mexico,* 24.

90. Nostrand, *Hispano Homeland,* 121.

91. Meinig, *Southwest,* 58.

92. Not all such activity was confined to the southwestern part of the territory. In 1904, stock raisers from the Chama Valley of Rio Arriba County in the north began to depasture flocks of sheep on lands long used by the Eastern Navajo. See Richard F. Van Valkenburgh, "A Short History of the Navajo People," 265, in Horr, ed., *Navajo Indians III,* 201–267.

93. H. A. Hoover, *Early Days in the Mogollons* (El Paso: Texas Western Press, 1957), 20. See also Evan Z. Vogt, *Modern Homesteaders* (Cambridge, Mass.: Harvard University Press, 1955), 159.

94. "Page B. Otero: New Mexico's Pioneer Game Warden," *New Mexico Conservationist,* April 1929, 3–4, 28; Elliot Barker, "A History of the New Mexico Department of Game and Fish" (manuscript, New Mexico Dept. of Game and Fish, 1970), 3–15; Otero to Outdoor Life Publishing Company, May 25, 1905, *Letterbook 1905–6;* Otero to Huntington, Oct. 30, 1905, *Letterbook 1905–6.*

95. Page Otero, untitled report to Miguel A. Otero, for period March 23, 1903, to June 30, 1903, p. 4, in Elliott Barker Papers, uncat. collection, Box 21, Rio Grande Historical Collections, Las Cruces, N. Mex. There were at least one or two sportsmen in the Socorro area. Neill B. Field was a recreational hunter who kept a ranch at Burley and corresponded with Otero in 1905, and W. M. Borrowdale was a wealthy local businessman whose letters to Otero were unique in complaining about white poachers as well as Indians. For Field, see Otero to Field, Oct. 30, 1905, *Letterbook 1905–6,* and Hening, *George Curry,* 194. For Borrowdale's poaching complaint, see Borrowdale to Otero, Nov. 6, 1905, *Letterbook 1905–6.* Borrowdale was a druggist whose lavish dinner parties were covered in the society columns of the local press. See "1888 Items," *Magdalena (N. Mex.) Mountain Mail* 12 (11) Nov. 1, 1990, 3.

96. Territory of New Mexico, *Acts of the Legislative Assembly of the Territory of New Mexico, 34th Session, 1901,* 97.

97. Territory of New Mexico, *Acts of the Legislative Assembly of the Territory of New Mexico, 35th Session, 1903,* 79.

98. Page Otero, untitled report, 1903, 2, Elliot Barker Papers, Box 21. Not all of these were in southwestern New Mexico. In 1904, Otero visited Taos Pueblo, where he informed a visiting Apache "who was there with 37 of his tribe" that no hunting would be allowed. "They went back to their reservation the next day." Page Otero, "Annual Report of the Territorial Game Warden, 1904," 2–3, Elliot Barker Papers, Box 21.

99. Territory of New Mexico, Joint Memorial 9, *1905 Acts of the Legislative Assem-*

bly of the Territory of New Mexico (Santa Fe, 1905), 389–390; the memorial cited incidents the length of western New Mexico in Taos, San Juan, Sandoval, Rio Arriba, Valencia, Socorro, Sierra, Grant, and McKinley counties.

100. Otero to Perry, Nov. 11, 1905, *Letterbook 1905–6.*

101. Otero to Department of the Interior, Nov. 21, 1905, *Letterbook 1905–6,* 228.

102. Wooton, "The Range Problem in New Mexico," 20; deBuys, *Enchantment and Exploitation,* 224.

103. Wooton, "The Range Problem in New Mexico," 19, 23.

104. After a mild recovery from the slump of the 1880s, the livestock market improved until 1898, when it began to decline again. Stock prices bottomed out in 1904 and 1905, at half their 1898 level. U.S. Dept. of Agriculture, *The Western Range* (Washington, D.C.: Government Printing Office, 1936), 125; Wooton, "The Range Problem in New Mexico," 24. Wooton's data suggest the depression in cattle prices was not as severe as a 50 percent drop. He lists the peak year for cattle prices in the 1890s as 1899, when the average price per animal was $18.64; the lowest year thereafter was 1904, when prices hit $13.84 per head. Prices were slightly better in 1905, at $14.84. In 1906, prices again hit $17.

105. Charles F. Coan, *A History of New Mexico,* vol. 1 (New York: American Historical Society, 1925), 470–471.

106. Parish, *The Charles Ilfeld Company,* 178–179. Taking the territory as a whole, sheep numbers were in decline at this time, and cattle numbers were on the increase. P. W. Cockerill, "A Statistical History of Crop and Livestock Production in New Mexico," *Bulletin 438,* New Mexico Agricultural Experiment Station, May 1959, 6, 11, 13.

107. Parish, *The Charles Ilfeld Company,* 178–179; for fences in Datil, see Cleaveland, *No Life for a Lady,* 320.

108. Wooton, "The Range Problem in New Mexico," 34–35.

109. Otero to Fullerton, Nov. 18, 1905, *Letterbook 1905–6,* 202–203; "Lone Hunter Meets Indian," *Silver City Independent,* Nov. 25, 1913, 1.

110. Otero to Blake, Dec. 26, 1905, *Letterbook 1905–6,* 282.

111. Nostrand, *Hispano Homeland,* 121.

112. Ibid., 107.

113. Game Warden Otero was Hispanic, but he was one of "los ricos," the small cadre of elite Hispanos who were largely partners of ruling Anglos in this period. See Sara Deutsch, *No Separate Refuge: Culture, Class, and Gender on an Anglo-Hispanic Frontier in the American Southwest* (New York: Oxford University Press, 1987), 29.

114. For evidence of Anglo dominance in conservation, see the lists of "deputy game wardens" published in annual reports of the game warden's office between 1912 and 1920, in the records of the New Mexico Department of Game and Fish, New Mexico State Archives and Record Center. Deputy game wardens were local volunteers who had powers of arrest in enforcing the game code. The lists show an overwhelming preponderance of Anglo names.

115. Territory of New Mexico, *Acts of the Legislative Assembly of the Territory of New Mexico, 1905,* 31–34.

116. Chuck Hornung, *The Thin Gray Line: The New Mexico Mounted Police* (Fort Worth, Tex.: Western Heritage Press, 1971).

117. Otero to Perry, Nov. 11, 1905, *Letterbook 1905–6,* 191.

118. Otero to Fullerton, Nov. 8, 1905, *Letterbook 1905–6,* 185.

119. Otero to McClure, Nov. 17, 1905, *Letterbook 1905–6,* 194.

120. Fullerton to Otero, Nov. 16, 1905, *Letterbook 1905–6,* 223.

121. Ibid.

122. Gifford Pinchot, *Breaking New Ground* (1947; repr. Seattle: University of Washington Press, 1972), 84–86.

123. On a visit to southeastern Arizona in 1909, Aldo Leopold suggested that rangers build local support for game laws by arresting tourist hunters who occasionally hunted out of season or exceeded bag limits. Meine, *Aldo Leopold,* 102. At least one local complained that forest rangers would not arrest white poachers. See Borrowdale to Otero, Nov. 6, 1905, *Letterbook 1905–6,* 216.

124. William D. Rowley, *U.S. Forest Service Grazing and Rangelands: A History* (College Station: Texas A&M University Press, 1985), 47–63.

125. Patterson to McClure, Nov. 8, 1905, *Letterbook 1905–6,* 219.

126. Benjamin W. Kemp, "Early Days in the Gila Country," *New Mexico Conservationist,* 2, no. 2 (December 1928): 27.

127. Otero to Braevogel, Nov. 17, 1905, *Letterbook 1905–6,* 201; Otero to Fullerton, Nov. 18, 1905, *Letterbook 1905–6,* 202–203.

128. Otero to Fullerton, Nov. 18, 1905, *Letterbook 1905–6,* 202–203.

129. Otero to Department of the Interior, Nov. 21, 1905, *Letterbook 1905–6,* 228.

130. Ibid.

131. "Red Men Are Lawless," *Socorro Chieftain,* Nov. 26, 1905, p. 1.

132. Otero to Andrews, Dec. 23, 1905, *Letterbook 1905–6,* 265.

133. Otero to Ferran, Dec. 10, 1905, *Letterbook 1905–6,* 246.

134. Otero to Blake, Dec. 11, 1905, *Letterbook 1905–6,* 247.

135. Otero to Blake, Dec. 26, 1905, *Letterbook 1905–6,* 282. See also, in the same letterbook, Otero to Borrowdale, Dec. 23, 1905, 275; Otero to Laird, Dec. 23, 1905, 277; Otero to Kendall, Dec. 23, 1905, 279; Otero to Fullerton, Dec. 23, 1905, 266; Otero to Stewart, Dec. 23, 1905, 268; Otero to Hubbard, Dec. 23, 1905, 269.

136. The sources are a particular problem on this point. The most inflammatory rhetoric of the 1905 struggle appears in the letterbook from that year, rather than in Otero's annual report. The same may have been true of years following, especially as annual reports became increasingly meant for public reading, but it is impossible to know because the 1905–6 letterbook is the only one that survives. The annual reports vary in their content and tone when they discuss Indians, with some years being remarkably more strident than others.

137. "Navajos Slaughter Deer," *Socorro Chieftain,* Nov. 16, 1907, p. 1.

138. W. E. Griffin to R. C. McClure, Oct. 30, 1906, *Letterbook 1905–6,* 702; NMGFW, *First Report of the Game and Fish Warden for New Mexico, 1909–1911,* 23.

139. District Court, Socorro Co. Criminal Cases, Case Number 3001–3023, NMDGF.

140. "Navajos in for Thirty Days," *Socorro Chieftain,* Nov. 23, 1907, p. 1.

141. NMGFW, *Report of the Game and Fish Warden for New Mexico, 1909–1911,* 23; also "Elk for New Mexico Game Preserves," *Las Vegas (N. Mex.) Daily Optic,* May 18, 1911, TANM, Reel 188, Frame 1310.

142. "Were Fined $500 and Costs," *Socorro Chieftain,* Nov. 19, 1910, p. 1. But according to the *Daily Optic* of Las Vegas, Gable confiscated the deer heads and hides, as well as the Indians' guns. See "Elk for New Mexico Game Preserves," in note 141.

143. "Apache Hunters Get Wise," *Socorro Chieftain,* Nov. 5, 1910, p. 1.

144. "Butchery in the Mogollons," *Socorro Chieftain,* Nov. 12, 1910, p. 1.

145. "Indians Fined for Slaughtering Deer," *Silver City Independent,* Nov. 3, 1914, p. 1.

146. NMGFW, *Report of the Game and Fish Warden of New Mexico, 1915–16,* 115.

147. Kemp, "Early Days in the Gila Country," 27.

148. Ibid.

149. NMGFW, *Report of the Game and Fish Warden of New Mexico, 1915–1916,* 37.

150. Bailey, *Mammals of New Mexico,* 25.

151. Meine, *Aldo Leopold,* 151.

152. Vogt, *Modern Homesteaders,* 159, 197.

153. Ibid.

154. Ibid.

Chapter 4. Tourism and the Failing Forest

1. Elliot Barker to Steve Rogers, Jan. 5, 1976, "Concerned Sportsmen," Box 6, Elliot Barker Papers, Rio Grande Historical Collections, New Mexico State University, Las Cruces (hereafter "Concerned Sportsmen").

2. Rogers to Barker, Jan. 7, 1976, "Concerned Sportsmen."

3. "The Peoples Column—Says Barker Confuses the Issue," *Albuquerque Journal,* Jan. 7, 1976, clipping in "Concerned Sportsmen."

4. Edwin J. Merrick to James Stephenson, March 2, 1976, "Concerned Sportsmen."

5. Walter A. Snyder, "Changes in New Mexico's Mule Deer Population and Management," *Proceedings of the Western Association of State Game and Fish Commissioners* 56 (1976): 404–407.

6. Ibid.

7. Concerned Sportsmen for New Mexico, "Attention Deer Hunters—Look What's Happening to Your Deer," leaflet in "Concerned Sportsmen."

8. Snyder, "Changes in New Mexico's Mule Deer Population and Management."

9. "Commission Takes License Control," n.d., clipping in "Concerned Sportsmen."

10. Curt Meine, *Aldo Leopold: His Life and Work* (Madison: University of Wisconsin Press, 1988), 161.

11. "Hunting Notes," *Silver City Independent,* Oct. 7, 1913, p. 1.

12. Ibid. See also "Hunting Notes," *Silver City Independent,* Oct. 21, 1913, p. 4.

13. "Sportsmen's Association Organized Wednesday," *Silver City Independent,* Sept. 30, 1913, p. 4.

14. "Arousing Enthusiasm for Game Protection," *Silver City Independent,* Sept. 23, 1913, p. 4.

15. *Silver City Independent,* Oct. 14, 1913, p. 2.

16. New Mexico Game and Fish Warden (hereafter NMGFW), *First Report of the Game and Fish Warden for New Mexico, 1909–11* (Santa Fe, 1912), 1.

17. "Miles Burford Called by Death," *Silver City Enterprise,* Nov. 9, 1917, p. 1; "Miles Burford Claimed by Death," *Silver City Independent,* Nov. 13, 1917.

18. These were Colin Neblett, Theodore Carter, and Wayne MacVeigh Wilson, respectively. For Neblett, see "Personal," *Silver City Enterprise,* Oct. 31, 1913, p. 4; for Carter, "Three New Buicks," *Silver City Independent,* Oct. 21, 1913, p. 4; for Wilson, see "Ellen Dickson Wilson," *Silver City Independent,* Jan. 27, 1914; "Go on Hunting Trip," *Silver City Enterprise,* Nov. 5, 1915.

19. "Sportsmen's Association Organized Wednesday," *Silver City Independent,* Sept. 30, 1913, p. 4.

20. Among the party were F. P. Jones, vice president of the American National Bank, and W. S. Cox, owner of W. S. Cox Incorporated, or The Busy Store, one of Silver City's better-known mercantile outlets. For Jones, see American National Bank advertisement in *Silver City Independent,* Nov. 3, 1914, p. 1. For Cox, see advertisements for The Busy Store in *Silver City Independent,* Nov. 18, 1913, p. 1, and Nov. 24, 1914, p. 1.

21. "Lone Hunter Meets Indian. Both Were Surprised and the Indian the Worst Skeered—Maybe," *Silver City Independent,* Nov. 25, 1913, p. 1.

22. "Combination to Wipe Out Predatory Animals," *Silver City Independent,* Dec. 7, 1915, p. 2.

23. Aldo Leopold, "Forestry and Game Conservation," *Journal of Forestry* 16, no. 4 (April 1918): 406–407. The quote is from p. 407.

24. Aldo Leopold, "The National Forests: The Last Free Hunting Grounds of the Nation," *Journal of Forestry* 17, no. 2 (February 1919): 152–153.

25. Aldo Leopold, "The Wilderness and Its Place in Forest Recreational Policy," *Journal of Forestry* 19, no. 7 (November 1921): 719–720.

26. Ibid., 721.

27. Robert H. Stewart, "Investigations of Big Game and Ranges: Historical Background of the Black Range," W-75-R-9, July 30, 1962, 18, in Records of the New Mexico Dept. of Game and Fish, File 146, New Mexico State Archives and Record Center, Santa Fe.

28. Roderick Nash, *Wilderness and the American Mind,* 3rd ed. (New Haven: Yale University Press, 1967), 184–185.

29. Leopold, "The Wilderness and Its Place in Forest Recreational Policy," 721.

30. Editorial, *New Mexico Conservationist* 1, no. 1 (September 1927): 8–9.

31. Ibid., 8.

32. Stewart, "Investigations of Big Game and Ranges," 8.

33. Ibid. For a review of the historical treatment of the Kaibab deer irruption, see Thomas Dunlap, "That Kaibab Myth," *Journal of Forest History* 32 (April 1988): 60–68.

34. Graham Caughley suggests it was the sudden availability of deer forage stemming from the removal of domestic livestock from the Kaibab. Graham Caughley, "Eruption of Ungulate Populations, with Emphasis on Himalayan Thar in New Zealand," *Ecology* 5 (1970): 53–72.

35. At the same time, the ponderosa or yellow pine forests on the higher slopes were encroaching upon the piñon-juniper forests. Aldo Leopold, "Grass, Brush, Tim-

ber, and Fire in Southern Arizona," *Journal of Forestry,* 22, no. 6 (October 1924): 2, 10, and "Pineries and Deer on the Gila," *New Mexico Conservationist* 1, no. 3 (March 1928): 3.

36. Leopold, "Grass, Brush, Timber." Although it concerns lower elevations, see also John York and William Dick-Peddie, "Vegetation Changes in Southern New Mexico During the Past Hundred Years," in William G. McGinnies and Bram J. Goldman, eds., *Arid Lands in Perspective* (Tucson: University of Arizona Press, 1969), 155–166.

37. J. Stokley Ligon, *Wild Life of New Mexico: Its Conservation and Management* (Santa Fe: State Game Commission and Department of Game and Fish, 1927), 185–189. In addition, although cattle in the region stayed along the watercourses and grazed them bare, they seldom ventured away from water to the higher mesas. There, some grasses still grew, and allowed mule deer to have necessary limited grazing in the early part of the year, when new growth had yet to begin on the brush. Leopold noted this phenomenon especially in the Gila Forest allotments of the GOS ranch, which included part of Black Canyon. Leopold, "Grass, Brush, Timber," 5.

38. Fred Winn, "The West Fork of the Gila River," *Science* 64 (July 2, 1926): 16–17; Aldo Leopold, "Erosion as a Menace to the Social and Economic Future of the Southwest," *Journal of Forestry* 44, no. 9 (September 1946): 627–633.

39. Leopold, "Grass, Brush, Timber," 2–3.

40. There has been a long debate among historians and geographers over the causes of the increased erosion of this period. Some argue that livestock grazing was the cause, and others that principal responsibility for the phenomenon lies with increased rainfall in the early twentieth century. I avoid taking sides in this dispute. Whichever was most responsible, livestock grazing certainly exacerbated the effects of rainfall, and only livestock grazing could be responsible for practically eliminating many grasses in heavily grazed localities. Luna B. Leopold, "Vegetation of Southwestern Watersheds in the Nineteenth Century," *Geographical Review* 41 (1951): 295–316; William M. Denevan, "Livestock Numbers in Nineteenth-Century New Mexico, and the Problem of Gullying in the Southwest," *Annals of the Association of American Geographers* 57 (1967): 691–703.

41. With a regime on the Gila Forest which forbade the opening of new roads and the use of motorized transport on old ones, Black Canyon became more isolated. By itself, this regulation practically eliminated hunting between Black Canyon and the Sapillo River and severely reduced it in Black Canyon. Stewart, "Investigations of Big Game and Ranges," 19.

42. Keith L. Bryant, Jr., "The Atchison, Topeka and Santa Fe Railway and the Development of the Taos and Santa Fe Art Colonies," *Western Historical Quarterly* 9 (October 1978): 437–453.

43. "The most salient characteristic of the art produced by Taos and Santa Fe artists and writers is the almost total lack of reference to or description of the immediate social reality within which it was created. . . . Rather than an arid, difficult environment governed by a pervasive system of stratified ethnic or race relations, New Mexico became an enchanted landscape populated by mystical tribes of noble savages." Sylvia Rodriguez, "Land, Water, and Ethnic Identity in Taos," in Charles L. Briggs and John R. Van Ness, eds., *Land, Water, and Culture: New Perspectives on Hispanic Land Grants* (Albuquerque: University of New Mexico, 1987), 344–345.

44. Dean MacCannell, "Reconstructed Ethnicity: Tourism and Cultural Identity in Third World Communities," *Annals of Tourism Research* 11 (1984): 377. I rely in the following argument on Rodriguez, "Land, Water, and Ethnic Identity in Taos," 313–403, esp. 342–347.

45. Elliot S. Barker, "A History of the New Mexico Department of Game and Fish," unpublished manuscript, New Mexico Department of Game and Fish, 1970, 44–45; also Elliot S. Barker, "History of New Mexico Game Protective Association," chapter three, 3, unpublished manuscript in Elliot Barker Papers, Box 20, Rio Grande Historical Collections, New Mexico State University, Las Cruces.

46. I am drawing again on Dean MacCannell's concept of "reconstructed ethnicity." Ethnicity here is plastic, with neither tourists, nor hawkers of tourism, nor local people exerting complete control over its formulation. Rather, all three groups contribute to the shaping process, with tourism and tourists representing a dominant, potentially overwhelming force in daily life. "Conforming to the requirements of being a living tourist attraction becomes a problem affecting every detail of your life. . . . Any deviation can be read as a political gesture that produces conflict not between groups but within the group." MacCannell, "Reconstructed Ethnicity," 389.

47. Ligon, quoted in Stewart, "Investigations of Big Game and Ranges," 10. I rely heavily on Stewart in the section that follows. Stewart's report was part of a Game and Fish Department effort to analyze the curious history of Black Canyon as a way of formulating policy responses to its changing ecology. Where possible, I have cited Forest Service materials in addition to Stewart. Most of the original Game and Fish Department sources, however, were apparently discarded. Writing in 1962, Stewart had access to many documents which have since vanished.

48. Quoted in Stewart, "Investigations of Big Game and Ranges," 10. See also "Memorandum on Examination of Deer Range on Gila National Forest," May 13, 1929, RG 95, Records of the Forest Service, Division of Wildlife Management, General Correspondence, W. Management, Box 3, Folder W-Management R-3 "Deer 1929 Examination of Deer Range on the Gila, Crook, and Coronado NFs," National Archives, Suitland, Md.

49. The Diamond Bar and GOS ranches held grazing allotments in Black Canyon. The Diamond Bar allotment did not show significant evidence of overgrazing by cattle. The GOS ranch, which had the highest cattle numbers in Black Canyon, immediately removed 3,000 of their 4,800 cattle from the allotment. "Memorandum on Examination of Deer Range on Gila National Forest," 3. Also Robert D. Baker, Robert S. Maxwell, Victor H. Treat, and Henry C. Dethloff, *Timeless Heritage: A History of the Forest Service in the Southwest* (College Station, Tex.: USDA, 1988), 50.

50. Quoted in Stewart, "Investigations of Big Game and Ranges," 10, 13.

51. Ibid., 13–14.

52. Ibid., 20–21.

53. Anonymous to R. P. Holland, Oct. 1, 1933; also Stuart to Holland, Oct. 9, 1933, RG 95, Records of the Forest Service, Division of Wildlife Management, General Correspondence, 1914–1950, W. Cooperation, Box 3, Folder "W. Cooperation R-3 State of New Mexico 1926–1936."

54. Stewart, "Investigations of Big Game and Ranges," 23.

55. Ibid., 24.

56. Barker, "A History of the New Mexico Department of Game and Fish," 100. For reference to sportsmen's animosity to the hunt in general, see Shoemaker to White, Feb. 4, 1936, 2, RG 95, Division of Wildlife Management, General Correspondence, W. Cooperation, Region 3, Box 3, Folder "W. Cooperation R-3 General 1915–1936."

57. With check stations on all access roads, the Game Department kept careful count of deer taken out of the canyon during the hunt. Estimating that another 500 deer were killed but lost in the brush, the total kill was put at 2,833 deer. Stewart, "Investigations of Big Game and Ranges," 22–23.

58. Barker, "A History of the New Mexico Department of Game and Fish," 100.

59. Stewart, "Investigations of Big Game and Ranges," 24.

60. Ibid., 24.

61. Shoemaker to White, Feb. 4, 1936, 2; Stewart, "Investigations of Big Game and Ranges," 29.

62. Stewart, "Investigations of Big Game and Ranges," 27.

63. In 1936, J. Stokley Ligon recommended re-establishing Black Canyon refuge because "the comparative scarcity of both deer and turkey necessitates protection to assist in restoration of normal deer hunting in that locality." Stewart, "Investigations of Big Game and Ranges," 27.

64. Ibid., 30.

65. D. A. Shoemaker, "Memorandum for Regional Forester," June 1, 1934, 3, RG 95, Division of Wildlife Management, General Correspondence, 1914–50, W. Cooperation, Box 3, Folder "W. Cooperation R-3 State of New Mexico 1926–1936."

66. Barker, "A History of the New Mexico Department of Game and Fish," 120.

67. Stewart, "Investigations of Big Game and Ranges," 34.

68. Ibid., 1, 38.

69. Ibid., 41.

70. For the following analysis, I draw on the work of George Gruell, who was commissioned by the Forest Service to evaluate the reasons for deer decline over much of the intermountain West. Although Gruell's study area was restricted to Montana, Wyoming, Idaho, Nevada, and Utah, the processes he described therein are so similar to those in the Gila Forest and in parts of Arizona and New Mexico that I conclude that they were probably identical. George E. Gruell, *Post-1900 Mule Deer Irruptions in the Intermountain West: Principle Cause and Influences* (Ogden, Utah: USDA–Forest Service, 1986).

71. Snyder, "Changes in New Mexico's Mule Deer Population and Management," 404–407.

72. For a sampling of the anguish prevalent in conservationist circles at the time, see: Guy E. Connolly, "Trends in Populations and Harvests," in Olof C. Wallmo, ed., *Mule and Black-Tailed Deer of North America* (Lincoln: University of Nebraska Press, 1981), 225–243; also the reports of various state wildlife managers on deer population trends in *Proceedings of the Western Association of Game and Fish Commissioners* 56 (1976).

73. Ira Gabrielson, *Report to the State Game Commission of New Mexico* (Santa Fe: New Mexico Department of Game and Fish, Nov. 21, 1949), 23.

74. "The Peoples Column—Says Barker Confuses the Issue," *Albuquerque Journal,* Jan. 7, 1976, clipping in "Concerned Sportsmen."

Chapter 5. Blackfeet and Boundaries

1. "Annual Report—Glacier National Park Season of 1928," *Superintendent's Annual Report 1928* (hereafter *SAR*), Ruhle Library, Glacier National Park, West Glacier, Mont. A very similar discussion found its way into *SAR 1929*. Unless otherwise indicated, all annual reports are in Ruhle Library, Glacier National Park, West Glacier, Mont.

2. John C. Ewers, *The Blackfeet: Raiders on the Northwestern Plains* (Norman: University of Oklahoma Press, 1958), 5, 284.

3. *U.S. Statutes at Large*, S. 2777, 1910, 61st Congress, Sess. II, Ch. 226, p. 354.

4. John C. Ewers, *The Horse in Blackfoot Indian Culture, with Comparative Material from Other Western Tribes* (1955; repr. Washington, D.C.: Smithsonian Institution Press, 1985), 163.

5. Ewers, *The Blackfeet*, 207–223.

6. Ibid., 236–253; Michael F. Foley, "An Historical Analysis of the Administration of the Blackfeet Indian Reservation by the United States, 1855–1950," Indian Claims Commission Docket 2790 [1974], 1–31.

7. Foley, 70–82, 103–106; Christopher S. Ashby, "The Blackfeet Agreement of 1895 and Glacier National Park: A Case History" (M.S. thesis, University of Montana, 1995), 14.

8. Ewers, *The Blackfeet*, 290, 304; Foley, "Historical Analysis," 70–71.

9. Ewers, *The Blackfeet*, 313; Foley, "Historical Analysis," 147, 166.

10. U.S. Senate, Senate Document No. 118, 54th Congress, 1st Session, "Letter from the Secretary of the Interior, Transmitting an Agreement Made and Concluded Sept. 26, 1895, with the Indians of the Blackfeet Reservation," 13. See also Grinnell's comment in same source, 17–18.

11. For pointing out many of the sources below, I am indebted to Mark David Spence, whose manuscript "Crowning the Continent: The American Wilderness Ideal and Blackfeet Exclusion from Glacier National Park" (1995, copy in author's possession) is the most synthetic and readable of sources on the subject of historic Blackfeet relationships to the park. The most recent in-depth ethnographic study of the issue of Blackfeet origins is in Brian Reeves and Sandy Peacock, " 'Our Mountains Are Our Pillows': An Ethnographic Overview of Glacier National Park," 2 vols., submitted to the National Park Service, Rocky Mountain Regional Office, Denver, Colo., vol. 1, 75–83. Reeves and Peacock convincingly disprove the contention of John C. Ewers and others that the Blackfeet migrated from the east. See Ewers, *The Blackfeet;* Oscar Lewis, "The Effects of White Contact Upon Blackfoot Culture with Special Reference to the Role of the Fur Trade" (1942; repr. Seattle: University of Washington Press, 1966), 7–15. Reeves and Peacock's argument confirms the earlier hypothesis of some scholars that the Blackfeet tenure in northern Montana is far older. See, for example, Clark Wissler, *Material Culture of the Blackfoot Indians* (New York: American Museum of Natural History, 1910), 18.

12. Spence, "Crowning the Continent"; Blackfeet annual cycle is in Ewers, *The Horse in Blackfoot Indian Culture*, 123–129. For buffalo hunting in the pre-horse era, see James Willard Schultz and Jesse Louise Donaldson, *The Sun God's Children* (New York: Houghton Mifflin, 1930), 37–38; Walter McClintock, *The Blackfoot Beaver Bundle*

(Los Angeles: Southwest Museum Leaflets, no. 2, n.d.), 10. For vision quests, see Reeves and Peacock, " 'Our Mountains Are Our Pillows,' " 1:139–173; Bob Frauson, interview with Mary Murphy, March 30, 1982, Oral History Collection, Ruhle Library; for bundles, see Spence, 4–5; Reeves and Peacock, 1:154–161.

13. C. R. Wasem, "1962–1963 Winter Elk Range Condition Report, Management Recommendations, and Range Study Methods Now in Practice as Well as Planned for Future Use," unpublished manuscript, National Park Service, Glacier National Park, November 1963, 10–11. Ruhle Library/GNP File 599.7357/Elk-GNP. See also Vernon Bailey and Florence Merriam Bailey, *Wild Animals of Glacier National Park* (Washington, D.C.: Government Printing Office, 1918), 32; *SAR 1929,* 12.

14. Jack Holterman, *Place Names of Glacier/Waterton National Parks* (West Glacier, Mont.: Glacier Natural History Association, 1985), 115; Gordon Pouliot, "Angus Monroe, Blackfeet Cowboy and Guide," unpublished essay, Ruhle Library; Joe Fisher, interview with the author, St. Mary, Mont., June 24, 1992. Tapes of all interviews in possession of the author. Monroe's history is in James Willard Schultz, *Many Strange Characters: Montana Frontier Tales,* ed. Eugene Lee Silliman (Norman: University of Oklahoma Press, 1982), 126.

15. George Bird Grinnell, "The Crown of the Continent," *Century Magazine* 62 (1901): 669. Although this article was not published until 1901, Grinnell had written it ten years before. See Warren L. Hanna, *The Life and Times of James Willard Schultz* (Norman: University of Oklahoma Press, 1986), 172.

16. Senate Doc. No. 118, comments of Pollock, p. 12, and Little Plume, p. 14.

17. Senate Doc. No. 118, 19.

18. Ibid., 21.

19. *Laws, Resolutions and Memorials of the State of Montana Passed at the Third Regular Session of the Legislative Assembly* (Butte City, Mont.: Inter Mountain Publishing, 1893), 72–73. The ban was rescinded by the 1897 legislature. *Laws, Resolutions, and Memorials of the State of Montana Passed at the Fifth Regular Session of the Legislative Assembly* (Helena: State Publishing, 1897), 249–250.

20. For the probable connection between the Bannock case and the 1895 negotiations with the Blackfeet, see Kenneth Pitt, "The Ceded Strip: Blackfeet Treaty Rights in the 1980s," n.d., 47, Ruhle Library GNP, File 1/970.4/Pi.

21. Ibid., 2; for a description of White Calf's career, see Ewers, *The Blackfeet,* 285–286.

22. For cabins, tipis, and livestock see Walter McClintock, "Hunting and Winter Customs of the Blackfoot" (1937–38), Series II, Folder No. 21, pp. 6, 28, Walter McClintock Papers, Beinecke Rare Book and Manuscript Library, Yale University. For firewood, see McClintock, "Hunting in the Rockies, the Beaver Medicine, and the Ceremony of Adoption," Series II, Folder 16, p. 14, Walter McClintock Papers; for lodge poles, gathering vegetables see Walter McClintock, *Old Indian Trails* (1923; reprint, New York: Houghton Mifflin, 1992), 156, 160; Schultz and Donaldson, *The Sun God's Children,* 42–43.

23. McClintock, "Hunting and Winter Customs of the Blackfoot" (1937–38), 2, 6.

24. McClintock, "Hunting and Winter Customs of the Blackfoot," 6a, See also Walter McClintock, *The Old North Trail* (London: Macmillan, 1910), 44–47.

25. See the testimony of J. Willard Schultz, George Pablo, John Ground, and

Francis X. Guardipee in Court of Claims of the United States, No. E-427 *Blackfeet et al. Indians vs. the United States, Evidence for the Plaintiffs* (Washington, D.C.: Government Printing Office, 1929), 119–127.

26. Grinnell, "The Crown of the Continent," 669.

27. Spence, "Crowning the Continent," 14; Gerald W. Diettert, *Grinnell's Glacier: George Bird Grinnell and Glacier National Park* (Missoula, Mont.: Mountain Press Publishing, 1992), 57; quotation from Grinnell, "The Crown of the Continent," 660.

28. Diettert, *Grinnell's Glacier,* 61–92.

29. Ann Hyde, *An American Vision: Far Western Landscape and National Culture, 1820–1920* (New York: New York University Press, 1990), 283; Diettert, *Grinnell's Glacier,* 94.

30. Spence, "Crowning the Continent."

31. Diettert, *Grinnell's Glacier,* 101; Spence, "Crowning the Continent," 14–16; Grinnell, "Notes to R. S. Yard," Oct. 27, 1927, Grinnell Papers, Box 31, Folder 112; C. W. Buchholtz, *Man in Glacier* (West Glacier, Mont.: Glacier Natural History Association, 1976), 3; Ashby, "The Blackfeet Agreement of 1895," 5–6.

32. Mike Anderson, "Local Opposition to the Creation of Glacier National Park, 1907–1912," manuscript, University of Montana, 1971, copy in Ruhle Library; Diettert, *Grinnell's Glacier,* 88–89.

33. Anderson, "Local Opposition"; *U.S. Statutes at Large* S. 2777, 61st Congress, Sess. II, Ch. 226, 354.

34. Camp Fire Club of America to Logan, April 7, 1911, Record Group 79, Records of the National Park Service, General Records, Central Files 1907–1939, Box 23, Folder "Game Protection," National Archives, Washington, D.C. (hereafter "Game Protection").

35. Logan to Secretary of the Interior, Oct. 14, 1911, "Game Protection."

36. The quotation is from Galen to Secretary of the Interior, Nov. 24, 1913, "Game Protection." Shields often advised tourist hunters to cross into the park; see Chapman to Secretary of the Interior, Nov. 24, 1912; also testimony of W. H. Sills before Commissioner Hutchings Nov. 21, 1912; all in "Game Protection." Shields was not the only federal employee to indulge in illegal hunting. Three Forest Service employees, all of them deputy game wardens, were arrested for hunting in the park in the fall of 1913. Galen to Secretary of the Interior, Dec. 5, 1913, "Game Protection." Employees of the Indian Service were also accused of poaching. DeHart to Lane, Aug. 5, 1916; Albright to DeHart, Aug. 24, 1916; DeHart to Albright, Aug. 30, 1916; all in "Game Protection."

37. Galen to Vaught, Dec. 16, 1913, "Game Protection"; Galen to Secretary of the Interior, Jan. 3, 1914, "Game Protection."

38. Galen to Secretary of the Interior, Jan. 23, 1914, "Game Protection." Warner was a resident of Belton, at the park's western entrance. Galen to Secretary of the Interior, Jan. 26, 1914, "Game Protection."

39. Ucker to Chapman, Aug. 1, 1912, "Game Protection." There is other evidence of Indian resistance to game laws: in early 1912, rangers arrested two Indians for hunting deer within the park boundaries. Rangers reported: "They also trapped one martin and killed five or six squirrels and a blue-jay." Hutchings to Logan, Feb. 6, 1912, "Game Protection." Also Hutchings to Secretary of the Interior (telegram), Feb. 3, 1912, "Game Protection."

40. Because there was no evidence that they killed any game on their expedition, the Indians were released, having forfeited their guns as a penalty for carrying them within park boundaries. Other park service complaints of the era reflect attempts by the communities east of the park line to retain hunting prerogatives in the customary hunting grounds of the ceded strip. The Reclamation Service posted rules against hunting, carrying firearms, and gathering green timber in the park. Their employees— Indian and white—working on roads between St. Mary and McDermott were arrested on several occasions. See the two letters from Galen to the Secretary of the Interior, both dated Nov. 10, 1913, "Game Protection."

41. Chapman to Secretary of the Interior, Nov. 24, 1912, "Game Protection."

42. See clipping "Cannot Hunt in the Glacier Park," and "Glacier Park in State Control," which explains Judge Bourquin's decision in the case, in set of documents dated Jan. 20, 1913–Jan. 20, 1914, "Game Protection."

43. Ucker to Chapman, June 18, 1912, "Game Protection"; Chapman to Secretary of the Interior, Aug. 28, 1912, "Game Protection."

44. Ucker to Assistant Attorney-General, Sept. 14, 1910; Ucker to Logan, Sept. 19, 1910; Ford to Logan, Oct. 17, 1911; Logan to Secretary of the Interior (telegram), Oct. 23, 1911; Acting Secretary to Logan (telegram), Oct. 24, 1911; Ucker to Galen, Feb. 20, 1913; Galen to Secretary of the Interior, Jan. 4, 1914; all in "Game Protection."

45. See West to Secretary of the Interior, Sept. 20, 1915, 2; West to DeHart, Nov. 4, 1915.

46. Landowner hunting and fishing was still subject to the laws of the state of Montana. The Interior Department considered fencing the park boundary where it adjoined private lands in order to prevent wildlife from crossing out of the park and into private property. "Memorandum on the Rights of an Individual Owning Property in the Glacier National Park, Montana, to Dispose of the Same, and to Hunt and Fish Thereon," "Game Protection." Authorities legalized fishing in the park a short time after its creation, although all fishers had to apply to the park superintendent's office for a license. See Clements to Chief Clerk, Sept. 9, 1910; Acting Secretary to Krausmann, September 1910, both in "Game Protection."

47. Acting Secretary to Geddes, Sept. 9, 1910, "Game Protection."

48. The first quotation is from Schoenberger to Secretary of the Interior, Sept. 21, 1910; second quotation from Schoenberger to Secretary of the Interior, October [illegible], 1911, "Game Protection."

49. DeHart to Lane, Sept. 28, 1915, "Game Protection."

50. "Memorandum," OWL to CDM, Dec. 6, 1915, "Game Protection."

51. Sullivan to Evans, Jan. 4, 1916; DeHart to Sweeney, Dec. 15, 1915; Mather to DeHart, Dec. 30, 1915, "Game Protection." Quotation in Ralston to Secretary of the Interior, Jan. 22, 1916, "Game Protection."

52. *SAR 1917,* 19.

53. See *SAR 1917,* 18; *SAR 1930,* 12.

54. The other was released. Assistant Secretary of the Interior to Hutchings, Feb. 17, 1912, "Game Protection."

55. LaBreche to Lane, Nov. 1, 1915, RG 75, Records of the Bureau of Indian Affairs, Central Classified Files, 1907–1939, Blackfeet 115 119292–1915, National Archives, Washington, D.C.

56. Mather to Solicitor, Sept. 4, 1915, "Game Protection."

57. West to Secretary of the Interior, Jan. 4, 1916, "Game Protection."

58. Sweeney to Ralston, Jan. 6, 1916, "Game Protection."

59. Hutchings to Secretary of the Interior, May 17, 1912; Acting Secretary of Commerce and Labor to Secretary of the Interior, May 29, 1912, both in "Game Protection."

60. Growing hay to feed deer is in Hutchings to Secretary of the Interior, March 1, 1912; importing elk is in Assistant Secretary to Hutchings, March 22, 1912; Hutchings to Secretary of the Interior, March 23, 1912; all in "Game Protection."

61. Galen to Secretary of the Interior, Jan. 2, 1913; Galen to Secretary of the Interior, Feb. 13, 1913; "Game Protection."

62. Seton to Secretary of the Interior, Nov. 23, 1916, "Game Protection."

63. DeHart to Myers, Jan. 29, 1916, "Game Protection."

64. Roderick Nash, *Wilderness and the American Mind,* 3rd. ed. (New Haven: Yale University Press, 1982), 111; Alfred Runte, *Trains of Discovery: Western Railroads and the National Parks* (Flagstaff, Ariz.: Northland Press, 1984), 19– 35.

65. Hyde, *An American Vision,* 283; Ralph W. Hidy, Muriel E. Hidy, Roy V. Scott, and Don L. Hofsommer, *The Great Northern Railway: A History* (Boston: Harvard Business School Press, 1988), 124–125; Runte, *Trains of Discovery,* 36–59. See also Ann Regan, "The Blackfeet, the Bureaucrats, and Glacier National Park," paper presented at the Western History Association conference, Billings, Mont., October 1986, 1, copy in Ruhle Library.

66. Secretary of the Interior to Hutchings, March 9, 1912; Price to Secretary of the Interior, March 6, 1912; for strychnine, Galen to Secretary of the Interior, Feb. 4, 1913; all in "Game Protection."

67. Hutchings to Secretary of the Interior, April 4, 1912, "Game Protection."

68. Mathewson to Mather, Feb. 14, 1916; Mather to Ralston, Feb. 21, 1916, both in "Game Protection."

69. Assistant Secretary to Galen, Jan. 6, 1913, "Game Protection." The Interior Department refused to allow importation of more elk from Yellowstone until there was "absolute game protection in Glacier." See handwritten notation by Ucker on letter of Galen to Secretary of the Interior, Feb. 13, 1913, "Game Protection."

70. Ucker to Chapman, June 21, 1912, "Game Protection."

71. *SAR 1913,* 15.

72. Hyde, *An American Vision,* 284; Ann T. Walton, John C. Ewers, Royal B. Hassrick, *After the Buffalo Were Gone: The Louis Warren Hill, Sr., Collection of Indian Art* (St. Paul, Minn.: Northwest Area Foundation, 1985), 24–25; Regan, "The Blackfeet, the Bureaucrats, and Glacier National Park."

73. Goodwin to Superintendent, June 19, 1917, "Game Protection"; Griffith to Sells, July 2, 1917, "Game Protection."

74. For assurances of help from the Office of Indian Affairs, see Merritt to Griffin, July 9, 1917, and Merritt to Mather, Dec. 3, 1917, "Game Protection."

75. See Albright to Payne, Jan. 4, 1918, "Game Protection."

76. Merritt to Wadsworth, Jan. 19, 1918, "Game Protection."

77. Secretary of the Interior to Sells, n.d., letter received Jan. 17, 1918, "Game Protection"; see also Merritt to Mather, Jan. 29, 1918; DeHart to Myers, Dec. 26, 1917; Myers to Secretary of the Interior, Jan. 2, 1918; Dorrington to DeHart, Dec. 12, 1917; all in "Game Protection."

78. Memorandum of Cato Sells, n.d., 4 (sub-file "March 6, 1916, to Dec. 6, 1918"), "Game Protection."

79. Ibid., 5.

80. *SAR 1917,* 19–20.

81. *SAR 1920,* 17.

82. *SAR 1921,* 11.

83. Burke to Campbell, March 17, 1923, reel 42, folder 168, Grinnell Papers, Yale University Manuscripts and Archives, New Haven, Conn.

84. *SAR 1923,* 8.

85. *SAR 1924,* 13–14.

86. Little Chief to Walsh, Jan. 23, 1928, enclosing copy of explanation of petition, RG 75, Central Classified Files, 1907–39, Blackfeet 115, 7807–1928. Also Regan, "The Blackfeet, the Bureaucrats, and Glacier National Park," 10–11.

87. Regan, "The Blackfeet, the Bureaucrats, and Glacier National Park," 10–11.

88. See *Blackfeet et al. Nations vs. U.S.,* in U.S. Court of Claims, *Court of Claims Reports* (Washington, D.C.: Government Printing Office, 1936), 81: 101–142.

89. See *SAR 1927,* 8.

90. *SAR 1929,* 12.

91. Eakin to Director, Dec. 14, 1929, Box 225, Ruhle Library Archives (hereafter RLA).

92. *SAR 1930,* 11.

93. *SAR 1933,* 9; *SAR 1934,* n.p.; *SAR 1935,* 15–16.

94. Scoyen to Director, May 25, 1932, RLA, Folder 116–1. Note: The copy is dated 1932, but in fact it appears to have been a 1931 letter. It is attached to a letter concerning the same issues, dated April 22, 1932, which refers to a letter of May 26, 1931, "a copy of which is attached." See also Ashby, "The Blackfeet Agreement of 1895," 52–53.

95. Scoyen to Director, April 22, 1932, Folder 196–2, RLA.

96. Ibid.

97. E. C. Finney, "Solicitor's Opinion of Blackfeet Rights on Glacier Park Land, 1932," June 21, 1932, Ruhle Library.

98. Testimony is in U.S. Court of Claims, *Blackfeet et al. Nations vs. U.S.: Evidence for the Plaintiff* (Washington, D.C.: Government Printing Office, 1929). The court's decision is in *Blackfeet et al. Nations vs. U.S.,* in U.S. Court of Claims, *Court of Claims Reports,* 115. Also, see Ashby, "The Blackfeet Agreement of 1895," 49–51.

99. *SAR 1937,* 15. See also Adolph Murie, "Memorandum for Mr. Wright," July 31, 1935, Folder 189–2, RLA.

100. For end of predator killing, see *SAR 1937,* 15–16.

101. *SAR 1938,* 8.

Chapter 6. Erasing Boundaries, Saving the Range

1. Sweeney to Grinnell, Sept. 4, 1927, reel 42, folder 168, Grinnell Papers, Yale University Manuscripts and Archives.

2. For the contribution of hunters to conservation, see John F. Reiger, *American Sportsmen and the Origins of Conservation* (1975; repr. Norman, Oklahoma: University of Oklahoma Press, 1985).

3. John C. Ewers, *The Horse in Blackfoot Indian Culture, With Comparative Material from Other Western Tribes* (1955; repr. Washington D.C.: Smithsonian Institution Press, 1985), 220; Walter McClintock, "Hunting and Winter Customs of the Blackfoot," 1937–38, McClintock Papers, Series II, No. 21, p. 21, Beinecke Rare Book and Manuscript Library, Yale University.

4. Versions of this tale are widespread. See George Bird Grinnell, *Blackfoot Lodge Tales: The Story of a Prairie People* (1892; repr. Lincoln: University of Nebraska Press, 1962), 158; Clark Wissler and D. C. Duvall, *Mythology of the Blackfoot Indians* (New York: Anthropological Papers of the American Museum of Natural History, 1908), 27–29; Walter McClintock, *The Old North Trail* (London: Macmillan, 1910), 340–341; Percy Bullchild, *The Sun Came Down* (San Francisco: Harper and Row, 1985), 200–205. Grinnell, Wissler and Duvall, and McClintock relate that Napi let the pregnant elk go to ensure the elk's survival; Bullchild implies that she escaped.

5. Some moved south of the Blackfeet Reserve, to Dupuyer Creek. Jack Holterman, "Chippewa Crees in the Glacier Park Country" (monograph, n.d., Ruhle Library File 7/719.85/Ho; a published version is available from the Glacier Natural History Association, West Glacier, Mont.); also Floyd W. Sharrock and Susan R. Sharrock, "A History of the Cree Indian Territorial Expansion From the Hudson Bay Area to the Interior Saskatchewan and Missouri Plains," Indian Claims Commission, Docket 221b-191, pp. 161–168, in *Chippewa Indians VI*, ed. David Agee Horr (New York: Garland, 1974).

6. Holterman, "Chippewa Crees."

7. Jack Holterman, *Place Names of Glacier/Waterton National Parks* (West Glacier, Mont.: Glacier Natural History Association, 1985), 16; George Bird Grinnell, "The Crown of the Continent," *Century Magazine* 62 (1901): 661.

8. Holterman, "Chippewa Crees." See also Holterman's *Place Names of Glacier/Waterton National Parks*, 16.

9. Holterman, *Place Names*, 116.

10. U.S. Census Bureau, *Twelfth Census of the United States, Census Bulletin No. 33*, "Population of Montana by Minor Civil Divisions," Jan. 17, 1901, 5.

11. U.S. Census Bureau, *A Report of the Seventeenth Decennial Census of the United States, Census of Population: 1950* (Washington, D.C.: Government Printing Office), vol. 2, pt. 26, "Characteristics of the Population," Montana, 26–11.

12. Clarence Wagner, interview with the author, Babb, Mont., June 15, 1992. All tape recordings of interviews in possession of the author.

13. John C. Ewers, *The Blackfeet: Raiders of the Northwestern Plains* (Norman: University of Oklahoma Press, 1958), 326–327.

14. Merlin Gilham, interview with the author, East Glacier, Mont., June 25, 1992.

15. Gilham, interview; field notes, Teddy Burns, interview with the author, July 1, 1992.

16. Juanita McKee, interview with the author, St. Mary, Mont., June 17, 1992; Darren Kipp, personal communication, June 1992.

17. Joe Fisher, interview with the author, St. Mary, Mont., June 24, 1992; field notes, Teddy Burns, interview, July 1, 1992.

18. Joe Fisher, interview; Merlin Gilham, interview; Clarence Wagner, interview; field notes, Teddy Burns, interview, July 1, 1992.

19. Gary Hannon, interview, Babb, Mont., June 21, 1992.

20. Vernon Bailey and Florence Merriam Bailey, *Wild Animals of Glacier National Park* (Washington, D.C.: Government Printing Office, 1918), 32.

21. Ibid.; *Superintendent's Annual Report 1917* (hereafter *SAR*), 18–20; George Bird Grinnell, "Journal," July 22, 1911, Southwest Museum, Los Angeles, Folder 351 (my thanks to Mark Spence for this last reference).

22. Juanita McKee, interview; Clarence Wagner, interview.

23. Joe Fisher, interview.

24. Clarence Wagner, interview.

25. Juanita McKee, interview; Clarence Wagner, interview.

26. Martelle W. Trager, *National Parks of the Northwest* (New York: Dodd, Mead, 1939), 92.

27. Joe Fisher, interview.

28. See Bert Gildart, "The Capture of Joe Cosley," *Montana Magazine,* 57 (January 1983): 47–49.

29. For an excellent study of the folklore about another poacher, see Edward D. Ives, *George Magoon and the Down-East Game War: History, Folklore, and the Law* (Urbana: University of Illinois Press, 1988).

30. Gary Hannon, interview, Babb, Mont., June 21, 1992.

31. Clarence Wagner, interview; field notes, Teddy Burns interview, June 11, 1992. Gary Hannon tells a version of this story in which the protagonist is not Joe Cosley but Van Pelt, a reclusive miner who owned a claim in the park. Hannon, interview.

32. Gildart, "The Capture of Joe Cosley," 47.

33. Clarence Wagner, interview.

34. Ibid.

35. Ibid.

36. "Story Regarding Capture of Joe Cosley, Indian Poacher by Joseph F. Heimes, Park Ranger, Glacier National Park As Told By Heimes," Box 115, Folder 3, Ruhle Library Archives, Glacier National Park (hereafter RLA/GNP), West Glacier, Mont.

37. Clarence Wagner, interview.

38. Teddy Burns, personal communication, June 1992.

39. *SAR 1937,* 15.

40. *SAR 1942,* 7; for 1930 figure, see *SAR 1930,* 11.

41. Jack R. Nelson, "Relationships of Elk and Other Large Herbivores," 459, and Jack R. Nelson and Thomas A. Leege, "Nutritional Requirements and Food Habits," 325, in Jack Ward Thomas and Dale E. Toweill, eds., *Elk of North America: Ecology and Management* (Harrisburg, Pa.: Stackpole Books, 1982).

42. Fred B. Hodgson, "Memorandum for the Chief Ranger," Jan. 1, 1944, Folder 200–11, RLA/GNP.

43. See Hillory A. Folson, "Memorandum for Regional Directors," Feb. 22, 1944, Folder 200–11, RLA/GNP.

44. See Ray Wedga, "Walton District," n.d., Folder 200–11, RLA/GNP.

45. Superintendent to Chairman of the House Select Committee on Conservation of Wildlife Resources A. Willis Robertson, Aug. 3, 1944, Folder 200–11, RLA/GNP.

46. Ibid.

47. Glacier Park Ranger Station, "Annual Wildlife Report for North Half of the

Marias District, 1945–6," Oct. 9, 1946, Folder 200–12; "Annual Wildlife Report For Hudson Bay District—1946," Folder 200–12; both in RLA/GNP.

48. J. C. Holroyd to Superintendent, Jan. 19, 1947, Folder 200–12, RLA/GNP.

49. Press release 47–11–13 (no. 19), Folder 200–12, RLA/GNP.

50. J. W. Emmert, "Memorandum for the Regional Director," Dec. 1, 1948, Folder 200–13, RLA/GNP.

51. Press release 47–11–13 (no. 19), Folder 200–12, RLA/GNP.

52. "Glacier National Park 1950 Annual Wildlife Report," 5, Folder 200-14, RLA/GNP.

53. "Annual Wildlife Report for Glacier National Park," Sept. 10, 1950, to Sept. 20, 1951, 10, RLA/GNP.

54. Ibid., n.p.

55. "Glacier National Park 1953 Annual Wildlife Report," Folder 201–1, RLA/GNP.

56. See the following papers in Folder 201–1, RLA/GNP: "1953–54 St. Mary Elk Herd Reduction Plan," "Long Range Management Plan for Middle Fork Elk Herd," "1953–54 Reduction Plan for Middle Fork Elk Herd."

57. J. W. Emmert to Regional Director, memorandum, April 3, 1953, Folder 201–1 RLA/GNP.

58. Ibid.

59. Ibid.

60. J. W. Emmert to Regional Director, memorandum, Jan. 20, 1954, Folder 201–1, RLA/GNP.

61. "1954–55 St. Mary Elk Herd Management Plan," Folder 201–1, RLA/GNP.

62. Ibid.

63. Emmert to Regional Director, memorandum, Aug. 1, 1955, Folder 201-1, RLA/GNP.

64. "Long Range Management Plan for Eastern Glacier Wildlife and Range," Jan. 7, 1963, Folder 201–4, RLA/GNP.

65. Barnum to Chief Ranger, memorandum, March 21, 1957, Folder 201–2, RLA/GNP.

66. Adams to Libby, June 23, 1942; Libby to Adams, June 26, 1942, Folder 196–2, RLA/GNP; quotes from Lowell Adams, "Report on the Wildlife of the Blackfeet Indian Reservation," 3–4, Department of the Interior, Aug. 1, 1942, 3, Record Group 75, Bureau of Indian Affairs, Blackfeet Indian Agency, Decimal Correspondence 020–034, Box 7, Folder 031.0, Federal Archives and Record Center, Seattle.

67. "1954–55 St. Mary Elk Herd Management Plan," Folder 201–1, RLA/GNP.

68. Adams, "Report on the Wildlife of the Blackfeet Indian Reservation," 3.

69. "The situation would, no doubt, be improved if we could get the Blackfeet Indians to leave a corridor of at least half a mile between the boundary and the reservation." "The Annual Animal Census Report for Glacier National Park," Sept. 20, 1953–Sept. 20, 1955, Folder 201-1, RLA/GNP. There is now a "buffer zone" on the reservation; locals remain suspicious of it as a park attempt to seize control of the reservation boundary. See Juanita McKee, interview; Gary Hannon, interview.

70. Joe Fisher, interview.

71. See the interview with Bob Frauson, March 30, 1982, Oral History Collections, Ruhle Library, GNP, 32. Also Juanita McKee, interview.

72. Joseph to Regional Director, memorandum, Jan. 18, 1956, Folder 201-2, RLA/GNP.

73. Joseph to Regional Director, memorandum, Feb. 16, 1955, Folder 201-1, RLA/GNP.

Epilogue

1. *United States vs. Woodrow L. Kipp,* 369 F. Supp. 74 (1974).

2. Author's New Mexico field notes, Oct. 20, 1991; Oct. 29, 1991; Nov. 9, 1991.

3. House Select Committee on Conservation of Wildlife Resources, *Conservation of Wildlife: Hearings before the Select Committee on Conservation of Wildlife Resources,* 76th Cong., 3rd Sess., 1940, 3. See also the map of Forest Service holdings between pages 126 and 127.

4. Ibid., 3, 7–8.

5. Joseph Kalbfus, *Dr. Kalbfus' Book: A Sportsman's Experiences and Impressions of East and West* (Altoona, Pa.: Times Tribune Co., 1926), 289.

6. The U.S. Department of Agriculture rescinded G-20A in 1941 with a new regulation providing for cooperative agreements between the Forest Service and the states. See Michael J. Bean, *The Evolution of National Wildlife Law* (New York: Praeger, 1983), 138–139.

7. Ibid., 121–122, 195–196.

8. Samuel P. Hays, *Conservation and the Gospel of Efficiency: The Progressive Conservation Movement, 1890–1920* (1959; repr. New York: Atheneum, 1969), iii.

9. The alienation of Hispanic communal lands to the National Forest Service has attracted the attention of a number of historians and anthropologists. See Malcolm Ebright, *Landgrants and Lawsuits in Northern New Mexico* (Albuquerque: University of New Mexico Press, 1994); Charles L. Briggs and John R. Van Ness, eds., *Land, Water, and Culture: New Perspectives on Hispanic Land Grants* (Albuquerque: University of New Mexico Press, 1987); William deBuys, *Enchantment and Exploitation: The Life and Hard Times of a New Mexico Mountain Range* (Albuquerque: University of New Mexico Press, 1985).

10. My approach to rural unrest borrows from scholarship on the developing world. See James Scott, *Weapons of the Weak* (New Haven: Yale University Press, 1985). Also William Beinart and Colin Bundy, *Hidden Struggles in Rural South Africa* (Berkeley: University of California Press, 1987), 1–45. A work in western history which borrows from some of the same scholarly traditions is John Walton, *Western Times and Water Wars* (Berkeley: University of California Press, 1992).

11. William K. Wyant, *Westward in Eden: The Public Lands and the Conservation Movement* (Berkeley: University of California Press, 1982), 27.

12. "The West at War," *Newsweek* 126 (3) July 17, 1995: 24–28; "Unrest in the West," *Time* 146 (17) Oct. 23, 1995: 52–66.

13. For the Counties Movement, see Florence Williams, "Sagebrush Rebellion II," 130–135; for the Wise Use Movement, see T. H. Watkins, "Wise Use: Discouragements and Clarifications," 45–52; both in John D. Echeverria and Raymond Booth Eby, eds., *Let the People Judge: Wise Use and the Property Rights Movement* (Covelo, Calif.: Island Press, 1995).

14. "The West at War," 24–28; "At Montana Siege, F.B.I. Spurns Aid of Rightist

Groups," *New York Times*, April 26, 1996; "Surrender Is a Victory for a Strategy of Patience," *New York Times*, June 14, 1996.

15. Eric Foner, *Reconstruction: America's Unfinished Revolution, 1863–1877* (New York: Harper and Row, 1988).

16. See Barbara Novak, *Nature and Culture: American Landscape and Painting, 1825–1875* (New York: Oxford University Press, 1980); Angela Miller, *The Empire of the Eye: Landscape Representation and American Cultural Politics, 1825–1875* (Ithaca: Cornell University Press, 1993).

17. Gifford Pinchot, *The Fight for Conservation* (1910; repr. Seattle: University of Washington Press, 1967), 95.

18. Evan Z. Vogt, *Modern Homesteaders* (Cambridge, Mass.: Harvard University Press, 1955), 155.

19. This criticism is widespread. See Charles Wilkinson, *Crossing the Next Meridian: Land, Water, and the Future of the American West* (Washington, D.C.: Island Press, 1992), for a good discussion of the problem.

Index